6 Forensic Science Progress

Forensic Science Progress

Volume 6

Shot Range Determination

By K. Sellier

With 66 Figures

Springer-Verlag
Berlin Heidelberg New York
London Paris Tokyo
Hong Kong Barcelona Budapest

Author

Prof. Dr. Dr. Karl Sellier
Institut für Rechtsmedizin der Universität
Stiftsplatz 12, W-5300 Bonn, FRG

Editors-in-Chief

Prof. Dr. A. Maehly
Forensic Science Centre, 21 Divett Place
Adelaide 5000 S. A./Australia

Prof. Dr. R. L. Williams
Ardmore, 9 Meon Road, Boscombe East
Bournemouth BH7 6PN, England

This volume is a revised and translated version of:
K. Sellier, „Schußentfernungsbestimmung" Band 7 der Reihe „Arbeitsmethoden der med. u. naturwiss. Kriminalistik", Schmidt-Römhild-Verlag, Lübeck 1967 — with permission of the publisher.

ISBN-13: 978-3-642-76723-4 e-ISBN-13: 978-3-642-76721-0
DOI: 10.1007/ 978-3-642-76721-0

Library of Congress Catalog Card Number: 86-640073

Softcover reprint of the hardcover 1st edition 1991

Typesetting: Thomson Press (India) Ltd, New Delhi
2152/3020-543210 – Printed on acid-free paper

Editorial Board

Editorial

During the years 1962–1965 Interscience Publishers produced a four-volume series called "Methods in Forensic Science". Since then no major effort seems to have been made to review the progress in the rapidly expanding field of forensic science.

Our new series "Forensic Science Progress" represents a serious effort to take up a neglected venture. The series intends to cover all aspects of forensic science but does not include forensic medicine which is well represented in other publications.

The aim of the publisher and the board of editors is to produce contributions of high quality by leading scientists in the field of forensic science. Suggestions for such contributions from the forensic community at large are of course also very welcome.

The volumes will not be topic-oriented but will give balanced views on various aspects of the science. The editors believe that the forensic worker should be informed about all branches of the science even if he may very well be specialised in one or few of them. Ideally, contributions should be from 40–80 typewritten pages. Experimental details, except when not published previously, should be covered by citing the appropriate references. Polemic passages should be avoided but this does not include objective criticism.

The publisher has tried to choose an editorial board which is representative not only of various topics but also of the various geographical regions of the world.

Editors — Publisher

Table of Contents

Arsenal): fulminate of mercury 35%, potassium chlorate 15%, black powder 40%, glass powder 5% and 5% gum arabic as binding agent.

New problems arose in the following years as smokeless powder increasingly replace black powder. The brass cartridge cases, with the introduction of the new propellant, became brittle and useless because of the formation of an amalgam (an alloy of the copper and mercury). In the era of black powder, this effect did not appear because the fouling caused by the black powder protected the brass from mercury vapor. Because of the introduction of the smokeless powder, new primer compositions had to be sought so that problems with the brass of the cartridge could be overcome. In any case, fulminate of mercury was suspect since after a short time at temperatures over 85 °C it exploded and also when stored at slightly higher than normal temperatures it rapidly became unusable.

The primary compositions for the European civilian market in 1910 were approximately those set out in Table 1. Round the turn of the century the military priming ingredients from various countries had a largely similar composition (Table 2). The Russian and British priming ingredients conformed qualitatively to the composition of the German and Austrian, whereas the American were free of mercury. How far these mixtures were based on the experience and not out of chemical considerations is shown by the history of the composition No 42 from Frankford Arsenal which had been introduced as early as 1910 by the military. It consisted of potassium chlorate 47%, antimony sulfide 31% and sulfur 22%. In 1917, millions of rounds failed to fire because the raw materials decomposed after being mixed. Therefore the USA had to introduce a new priming composition which had been developed by Buell in 1911:

Table 1.

	Belgian	Eley	Kynoch	Mauser
Fulminate of mercury	42.9	41.1	39.3	9.1
Antimony sulfide	—	11.0	26.3	39.4
Potassium chlorate	52.3	27.9	34.2	45.5
Glass powder	—	19.0	—	3.0
Resin	4.8	—	—	—
Sulfur	—	—	—	3.0

Table 2. Priming compositions used by the military at the turn of the century

	Austria Hungary	Germany Composition 88	USA H-48(1898) for .30 Krag Cartridge
Fulminate of mercury	14	27	—
Antimony sulfide	33	29	25.1
Potassium chlorate	41	37	49.3
Glass powder	11	.7	16.6
Sulfur	—	—	8.7
Binding agent	.6	.6	—

Potassium chlorate 53%, lead thiocyanate 25%, antimony sulfide 12%, and TNT 5%.

This composition was used by the USA military under the designation FA70 in pistol cartridges and rifle ammunition right up to the 1950s.

Military primer ingredients must be capable of being stored for almost unlimited periods and production deviations affecting performance must be kept to a minimum. For these reasons, during the Second World War the USA as well as Germany did not switch from their proven mixtures for a long time and the Russians not at all. This was in spite of the fact that, after firing, the burnt residues had a strongly corresive effect on the barrels. The potassium chlorate above all caused the dreaded afterfire or after-rusting in the barrels.

The first corrosion free priming composition was brought on to the market by RWS (Rheinisch-Westfälische Sprengstoffwerke) as priming composition 1901. It consisted of fulminate of mercury 11%, barium peroxide 27% and TNT 7%.

This was the first time that the chlorate had been replaced by an alkaline component (barium peroxide). Only years later was RWS able to remove the poisonous and expensive fulminate of mercury from the primer ingredients. It was new chemical compounds which made this possible—tetracene and lead styphnate.

Then in 1928, a combination of these chemical compounds was put together in a new priming composition, SINOXID which was to be a prototype of this kind:

Lead styphnate	25–55%
Tetracene	0.5–5%
Barium nitrate	24–25%
Lead dioxide	5–10%
Calcium silicide	3–15%

From this the German Military Priming Composition 39/40 was developed as a replacement of the 88 Composition and it consisted as follows:

Lead styphnate	40%
Tetracene	3%
Sodium nitrate	42%
Lead dioxide	5%
Calcium silicide	10%

The exact composition was varied according its purpose, so, for example, glass powder was generally added to rim fire primers. Naturally, competing companies tried to circumvent the patents held by RWS by using copper and silver salts of tetracene or other easily disintegrating organic compounds.

Priming compositions which are analogs of SINOXID are today universally in use. Predominantly, lead styphnate was employed as the alkaline styphnate (easy and safe to produce), so, for example, in 1949 the King-Federal Cartridge Company produced the following composition for shot shells:

Lead styphnate	40%
Barium nitrate	42%

Antimony sulfide	1%
Nitrocellulose	6%
Tetracene	1%

Companies also tried to do without lead completely. So, as described in U.S. patent from 1935, McNutt (Winchester) replaced the styphnate with diazodinitrophenol, according to a patent from Brün (Winchester) the tetracene was replaced by hypophosphate:

Lead styphnate	33%
Calcium hypophosphate	7%
Lead nitrate	14%
Lead thiocyanate	10%
Barium nitrate	16%
Glass powder	20%

An interesting development seemed to open with the introduction of pyrophoric metals. Eschbach (Dynamit-Nobel), in 1936, introduced alloys from rare earth metals, e.g. "Mischmetall" (50% cerium, 40% lanthanum, 3% rare earths, 7% iron). Foote and Sahlim made further developments in 1964. A very useful mixture consisted, for example, of:

"Mischmetall"-Magnesium 80/20	50%
Barium nitrate	20%
Lead dioxide	10%
Zirconium powder	20%

The composition of military primer ingredients is shown in Table 3 and 4.
In deference to modern concern with the environment, investigation of shooting galleries were made. The results showed that in many cases the permissible level of lead in the air—varying with the number of shots per day—was exceeded. The ammunition industries therefore involved itself in the development of lead-free primer ingredients. This development was begun as early as the 1950s. So, for example, Kenney replaced the lead compounds with iron tricinate and iron

Table 3. Further military primers

	Austria	Germany 88	Engl. A.	Russ WW I WW II	H-48 1898	USA Fa 42 1910	FA70 1917–50
Fulminate of mercury	14	27	50	16.7	—	—	—
Potassium chlorate	41	37	25	55.5	49.3	47	53
Antimony sulfide	33	29	25	27.8	25.1	31	17
Lead thiocyanate	—	—	—	—	—	25	
Glass powder	11	7	—	—	16.6	—	—
Sulfur	—	—	—	—	8.7	22	—
TNT	—	—	—	—	—	—	5
Binding agent	0.6	0.6	—	—	—	—	—

Table 4. Modern military primer composition

	Germany 30/40	Germany NATO	Czech	FA675 ca. 1950	USA FA90 II. WK	FA956 5.56 and 7.62	Russia M43
Fulminate of Mercury		—	—	—	—	—	25
Lead styphnate	40	40	45	—	—	37	—
Lead thiocyanate	—	—	—	—	25	—	—
Tetracene	3	3	5	—	—	4	—
Antimony sulfide	—	—	20	—	12	15	37
Red phosphorus	—	—	—	18	—	—	—
Potassium chlorate	—	—	—	—	53	—	38
Barium nitrate	42	42	25	82	—	32	—
Nitropenta	—	—	—	—	10	5	—
Lead dioxide	5	5	—	—	—	—	—
Calcium silicide	10	10	—	—	—	—	—
Al-Mg mixture	—	—	5	—	—	7	—

hypophosphite. (U.S. patents—Remington 1954, 1955). Recently, Dynamit-Nobel brought a new lead-free priming composition (SINTOX) onto the market. This will replace SINOXID and contains only Zn and Ti salts. In addition, in these cartridges (jacket free) there is a cup at the base of the bullet, were the bullet core becomes visible, which hinders vaporization of the lead during a shot.

It is evident, that it is important to know, even before the beginning of an examination, what powder soot elements can be expected to be present in the target so that the appropriate method can be chosen. Fortunately the name of the primer of the 9 mm Parabellum cartridge from the Dynamit-Nobel Co. ("Geco") is stamped into base ("SINTOX" or "SX"), in contrast to the cartridges .38 special with partly jacketed bullets and the .38 Wad Cutter from the same manufacturer which cannot be distinguished outwardly from those with SINOXID primer. The latter is not to prevent lead from being discharged (the pure lead bullet does so abundantly), but rather to improve the internal ballistics. The 9 mm Parabellum cartridge from CCI company (USA) is supplied with either a normal or a SINTOX-similar primer.

To prevent a discharge of lead it is important, but not completely sufficient, to use a lead-free primer. The lead core of the bullet itself must be totally enclosed. So, for example, whereas in the 9 mm Parabellum-SINOXID cartridge the base of the bullet is uncovered, in the new SINTOX cartridge it is enclosed. With the above mentioned .38 special cartridge with a partly jacketed bullet and because of the exposed head of the bullet one finds in a close range shot not just Zn and Ti (from the primer) but also abundant Pb (from the bullet core). Lichtenberg (1985) investigated this with a systematic series of experiments.

The absolute amount of the priming ingredients for pistols and revolvers lies between 18 and 24 mg. A primer for the 9 mm Parabellum cartridge (Luger) produces a charge of 22 mg. One primer which is often used in gunshot cartridges produces a charge of 38 mg. With rapid development taking place, we can expect some distinct changes in the composition of primers. So, for example, .22 l.r.

ammunition produced in Germany has contained no antimony for sometime (despite this antimony can appear on the target in the case of a close range round as an alloyed part of the hardened lead).

2.1.1.2. Powder

As previously mentioned, following the bullet from the barrel, come unburned and partially burned powder grains which, with an intermediate range gunshot, hit the target with considerable impact and are compressed into it. There are numerous kinds of powder whose various characteristics must be made suitable for the bullet. The complicated inner ballistic processes will not be dealt with here (see, for example Cranz (1926), Kutterer (1959)). A selection of such powders is shown in Fig. 1a to d. The determination of the kind of powder grains in an intermediate range shot can say which powder was used and furthermore possibly the kind of cartridge when cartridges from different manufactures come into question.

2.1.1.3. Bullets and Cases

Revolver bullets consist mostly of lead hardened by being alloyed with a few % antimony. However, jacketed bullets are also produced, especially for Magnum

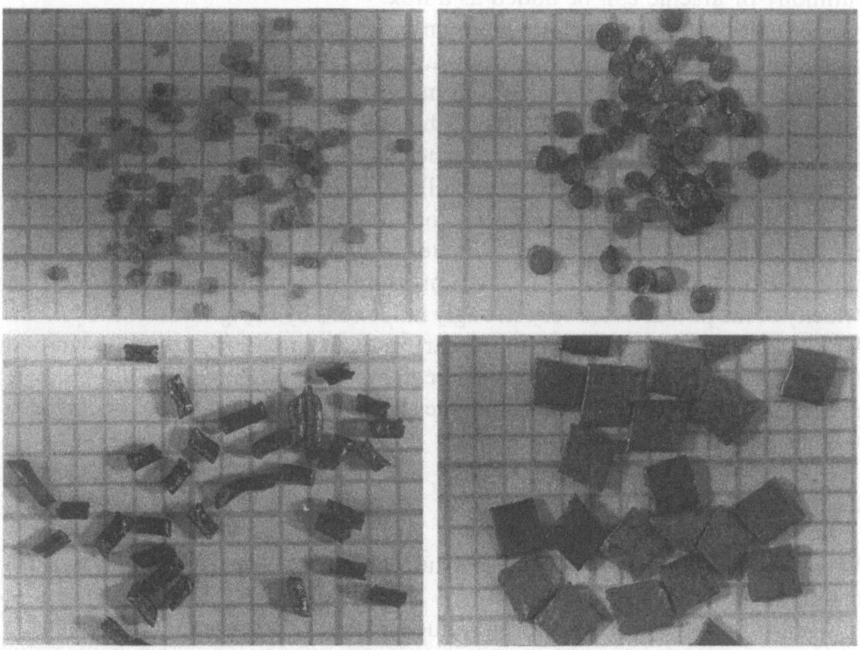

Fig. 1 a–d. Various kinds of smokeless powder. The shapes **a–d** and pretreatment of the individual kings give different burn out behavior

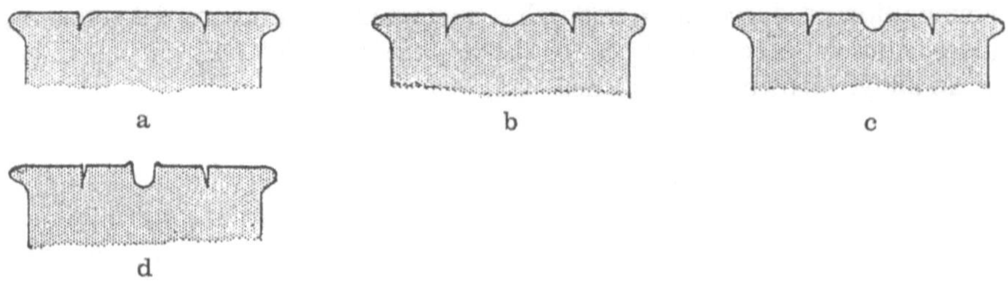

Fig. 2 a–d. Longitudinal sections through cartridge cases. These show the effect of the pressure on the primer. The imprints show (from Burrard, 1970). **a)** unfired cartridge; **b)** weak pressure; **c)** normal pressure; **d)** high pressure

cartridges, where the bullets reach a relatively high velocity. Normal lead bullets, because of their softness, would jump over the lands and not receive adequate spin.

Pistol cartridges (with the exception of small caliber cartridges) are loaded with full jacketed bullets, the jacket consisting of a copper-zinc alloy (tombac, brass) but can also be nickel-plated. This is also the case for rifle bullets. Bullets for semi-automatic pistols need a jacket because a soft bullet head would lead to feed jamming when used on repeat.

Shotgun pellets consists of hardened lead, to the outer layer of which antimony or arsenic can be added as a flux.

The cases consist of brass (in Germany with raw material shortages towards the end of the Second World War, even varnished steel was used). Cases only have relevance to shot range determinations in that the case head bears a head stamp which informs us of the brand and possibly the year of production. This will not be dealt with here and many reference works are available (Brandt and Hartmann, Pawlas, White and Munhall).

The shape of the firing pin impression in the primer at the base of the case provides information on the approximate amount of gas pressure during a shot, naturally only roughly (Fig. 2). This pressure is a standard for combustion velocity of the powder. It is very strongly influenced by the ignition of the primer. After long or improper storage of the cartridges, it can happen that the ignition charge can partly decompose or become damp. The powder then burns slowly and the shot is weak. This can sometimes be very important when coming to a judgement.

2.1.2 Shot Shells (Shotgun Cartridges)

A shotgun shell consists of a cylindrical hull containing instead of a single bullet, a charge of shot.

The hull consists of wound and glued paper, brass, aluminum, or plastic material with a head made from brass, or brass plated, or varnished steel.

They are 65 or 70 mm long. Thus measurement applies to the hull after it has been fired. When live, the hull (cartridge) is shorter due to the crimping or folding.

The length of the hull—and therefore its capacity—can be obtained from the designation on the cartridge, e.g. 12/70 or 16/65.

The propellant consists of mono- or di-basic nitrocellulose powder (black powder cartridges can still be obtained). The section containing the powder is sealed off from the charge of shot by a filler of felt, cork, or plastic. The closure is made with an over-the-shot wad of cardboard held in place by either rolled crimping of the rim of the shell or by star-shaped crimping of the mouth of the shell. The mass of the charge of 12 gauge cartridges is about 32 g (according to purpose this can be higher-up to 40 g- or less), in 16 gauge about 30 g, and in 20 gauge a maximum of 29 g.

Currently, two types of cartridge can be considered as normal. (see Figs. 3 and 4). Type 1 has a cardboard case, nitrocellulose powder, a felt wad and a charge mass of 32 to 36 g. This cartridge is used mainly by European hunters because of its good environmental characteristics. Type 2 has a plastic case, mono- or di-basic powder, a plastic cup-shaped wad, and a charge of shot weighting between 28 and 32 g.

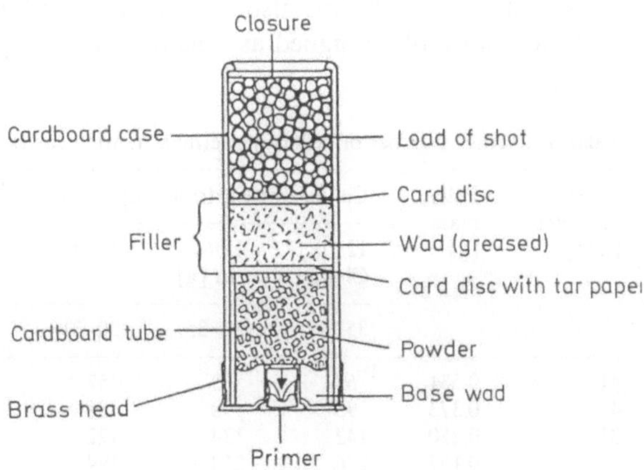

Fig. 3. Shot shell-Type 1 with cardboard hull and felt wad

Fig. 4. Modern shot shell-Type 2 with plastic case, plastic wad and -cup

The above-mentioned filler separating the powder and the charge of shot is of great importance. It has the task 1.) sealing the powder gases as tightly as possible against the charge of shot and 2.) because of its elastic properties accelerating the charge of shot "gently". These tasks are adequately fulfilled by a lubricated felt wad. In type 2 cartridge a plastic wad is used and this can have various forms. One form is shown in Fig. 4 There are always 3 elements: seal, elastic part (resilient spring), and a polyethylene cup. The charge of shot is completely surrounded by the cup. The wall of the cup is split (mostly 4 times) so that after exiting from the barrel it can open like wings and the shot column can come out without hindrance (see e.g. Fig. 57d).

Rubbing of the shot within the barrel is prevented by the cup and this gives these cartridges a better pattern because "flyers" hardly ever occur. Of importance for the pattern is that the use of such a cartridge strongly reduces the influence of the choke in that shot column is held together better.

Finally, the number of pellets in a cartridge of a particular caliber and length is important, e.g. in judging a partial pattern. These values are given in Table 5. For many years there have been shot shells with the caliber .22 for handguns and rifles, and for a short time also for .38 and .44 Magnum calibers (Fig. 5). These shells are normally designed as "snake-shot" or "rat-shot" and are used in the

Table 5. Average number of pellets in German hunting cartridges

Pellet diameter [mm]	Pellet mass [g]	Caliber- or Cartridge Identifier					
		12/70	12/65	16/70	16/65	20/70	20/65
		Charge mass in [g]					
		35–36	32–35	30–31	28–29	26–27	25–26
$4\frac{1}{2}$	0.531	67	63	57	54	50	48
4	0.373	95	90	81	76	71	68
$3\frac{1}{2}$	0.250	142	134	122	114	106	102
3	0.157	226	214	194	181	169	162
$2\frac{1}{2}$	0.091	390	368	335	313	291	280

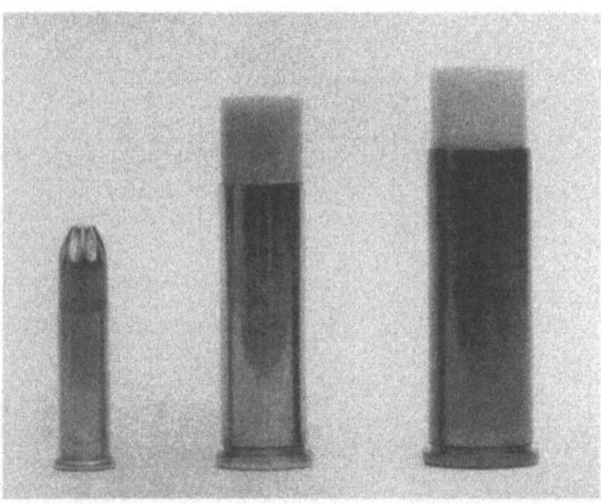

Fig. 5. Small caliber shot shells i.e. less than 20 gauge—from left to right .22; .38; .44 Magnum

Table 6. Compilation of shot shells with a caliber under 20

Producer	Cartridge	mass [g]	dia. [mm]	No. *1	v_0 [m/s]	shot *2	remarks
Fiocchi	Flobert 6 mm Einfachschrot	1.07	1.5	≈ 53	150	–	
RWS	Flobert 9 mm Einfachschrot	3.7	1.0	≈ 185	160	–	so called "dust"
Fiocchi	Flobert 9 mm	7.9	1.9	≈ 200	220	–	
RWS	.22 lr.	2.0	1.5	≈ 100	300 ± 20	–	
CCI Lewiston IDAHO	.22 Winchester Magnum	3.5	1.5	≈ 170	k:320 1:380	+	sh. ba.* 1. ba.**
Speer Lewiston	.38/.357	6.6	2	≈ 140		+	
Speer Lewiston	.44 Magnum	8.9	2	≈ 190		+	

* —Number of shot pellets
* 2—plastic case
* —sh. ba. = short barrel; ** 1. ba. = long barrel

control of vermin. As the ballistic data of these shot shells are little known, they are listed in Table 6.
Rem.: For the Flobert-shotshells it is possible, depending on the manufacture (RWS, Fiocchi, Sellier & Bellot) that there is a fluctuation in the v_0, the pellet diameter, and the number of pellets. For example, it is stated for the RWS-"Doppelschrot"-shell: m = 6.9 g and v_0 = 180 m/s.

For Flobert pistols and rifles with calibers of 6 and 9 mm, shot shells have also been available for a long time. They were certainly the model for the construction and introduction of shotshells in .22, .38, and .44 calibers.

2.2 Sequence of Events During Firing

When the firing pin strikes the base of the primer, the priming ingredients detonate producing a flame which travels through the flash hole or flash holes into the powder chamber. This causes the powder to ignite and produce gases to generate extremely high pressure which propels the charge of shot or bullet out of the barrel (Figs. 6 and 7). The powder never burns completely—it depends essentially on the powder of the primer—therefore burned and unburned powder particles emerge from the muzzle.

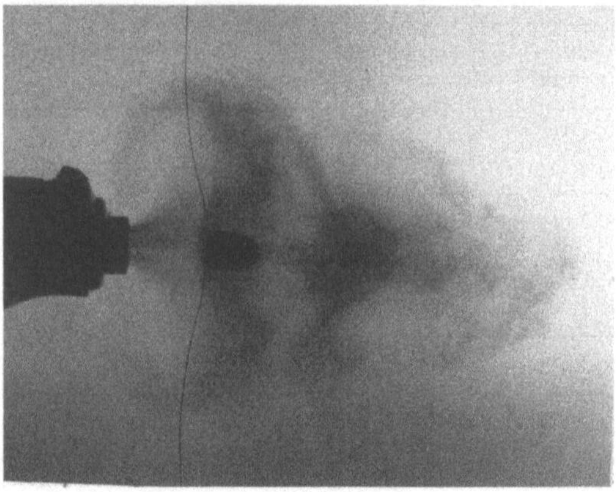

Fig. 6. Soot distribution in front of the barrel. Notice that some soot has emerged before the bullet (a 7.65 Browning). Notice further the shaping of the emerging of gas by the lands and the rifling

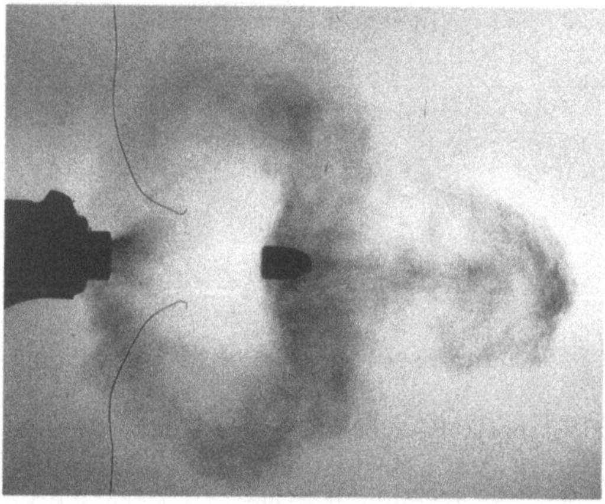

Fig. 7. As Fig. 6 but showing the distribution 200 μs later

Ideally the powder should oxidize to H_2O, CO_2 and nitrogen oxides (N_xO_y). However, carbon products poorer in oxygen, CO and C are present in the combustion gases in considerable quantities, e.g. CO up 50%. The unburnt carbon is in the form of extremely fine particles ("molecular") as soot. Its color, with a close range shot, is from gray to black in the vicinity of the hole caused by the shot. This soot, the incomplete combustion product from the powder is called powder soot and the blackish coloration powder-soot blackening in forensic ballistics (Fig. 8).

The amount of powder-soot on the target is proportional to the amount of powder in the cartridge: the larger the charge—and therefore the energy of the cartridge—the more soot is emitted from the muzzle. The pattern of the powder-soot blackening with reference to its distribution and intensity is also a function of the shot range.

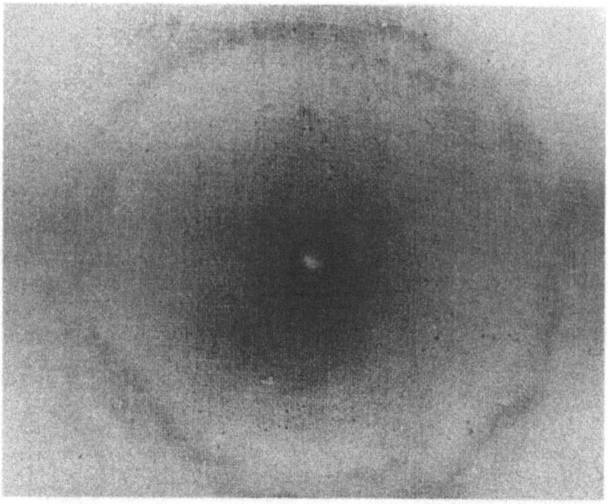

Fig. 8. A typical pattern of a nearer intermediate range shot. The range was 8.5 cm and the weapon a Walther PP 7.65 pistol. The intensively black ring around the entry hole is the bullet wipe ring from the bullet. This is surrounded by blackening due to soot and powder particles. The large-diameter ring is the pattern caused by a vortex ring (see Figs. 6, 7 and Fig. 9e). This ring only appears at short range and this depends on the weapon and the ammunition

This distribution can be roughly approximated to an exponential distribution

$$I = I(s) \cdot \exp(-a \cdot r),$$

where I(s) = the maximum intensity at the range s, and r = the distance from the center of the entry hole. With increasing range the density of the powder-soot pattern decreases whereas its diameter increases. When the same sort of cartridge is used, the powder soot distribution is dependent on the length of the weapon's barrel. A longer barrel keeps the soot more together, so that at the same range a long-barreled weapon produces a more intensive pattern than a short-barreled weapon at a closer range (Fig. 9).

These differences are particular obvious for weapons for the .22 l.r. It is well-known that these cartridges can be fired from long and short weapons (rifles, pistols and revolvers). The barrel length can vary from 65 cm (rifles) to 5 cm (2" revolver). The differences in the powder soot patterns for the same cartridges (.22) and the same shot range are very considerable (Figs. 10 and 11).

The distribution and intensity of the powder-soot picture is therefore dependent on the following factors:

1. The size of the powder charge (also the kind of powder), (Fig. 12)
2. The length of the weapons barrel
3. The shot range
4. The characteristics of the target materials (glass, textiles)

The powder grains which emerge from the muzzle at high velocity strike the target when the range is not too great. According to the hardness of the target, they are either forced into it or they drop from it (Figs. 13–16). The pattern diameter and the density of the particles (number/unit area) are a function of the shot range. One can very roughly say that the powder grains—similarly to shotgun pellets—spread out from the barrel in a cone the apex of which is the muzzle.

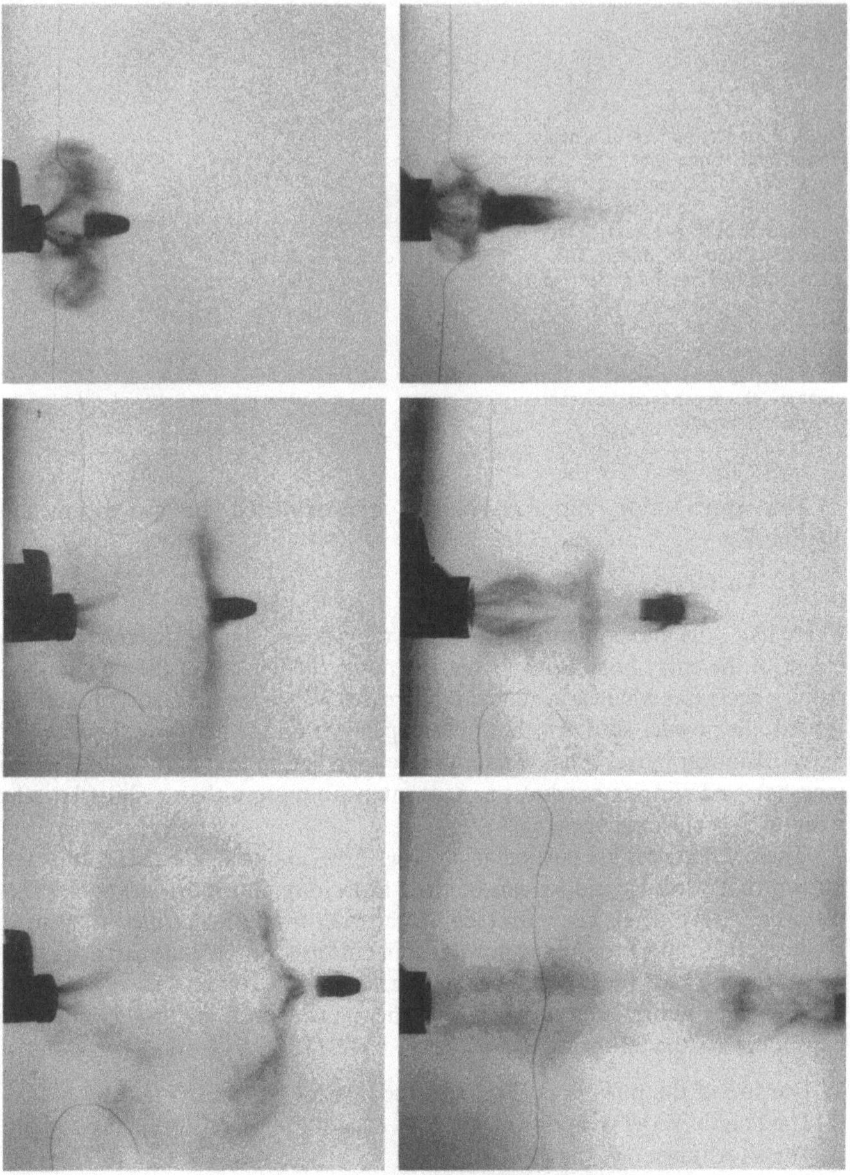

Fig. 9 a–f. Influence of the barrel length on the radial spread of soot. Ammunition .22 l.r.: **a–c** short barrel (Walther PP Sport pistol); **d–f** long barrel (small bore rifle)

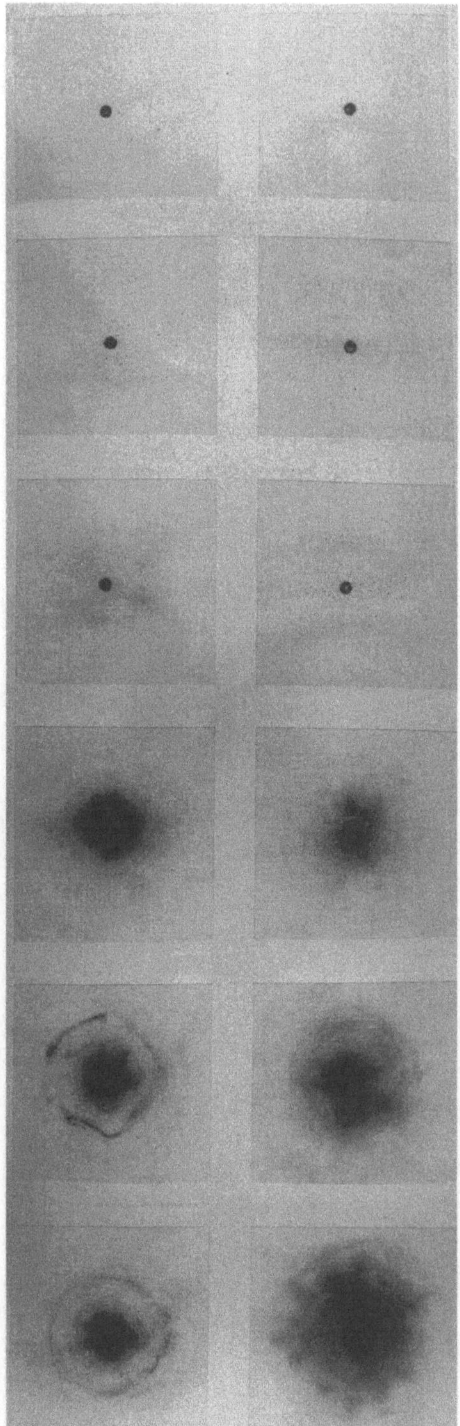

Fig. 10. Influence of barrel length on the soot pattern (the range of the shot decreases from top to bottom): on the left rifle, on the right pistol. Ammunition .22 l.r.

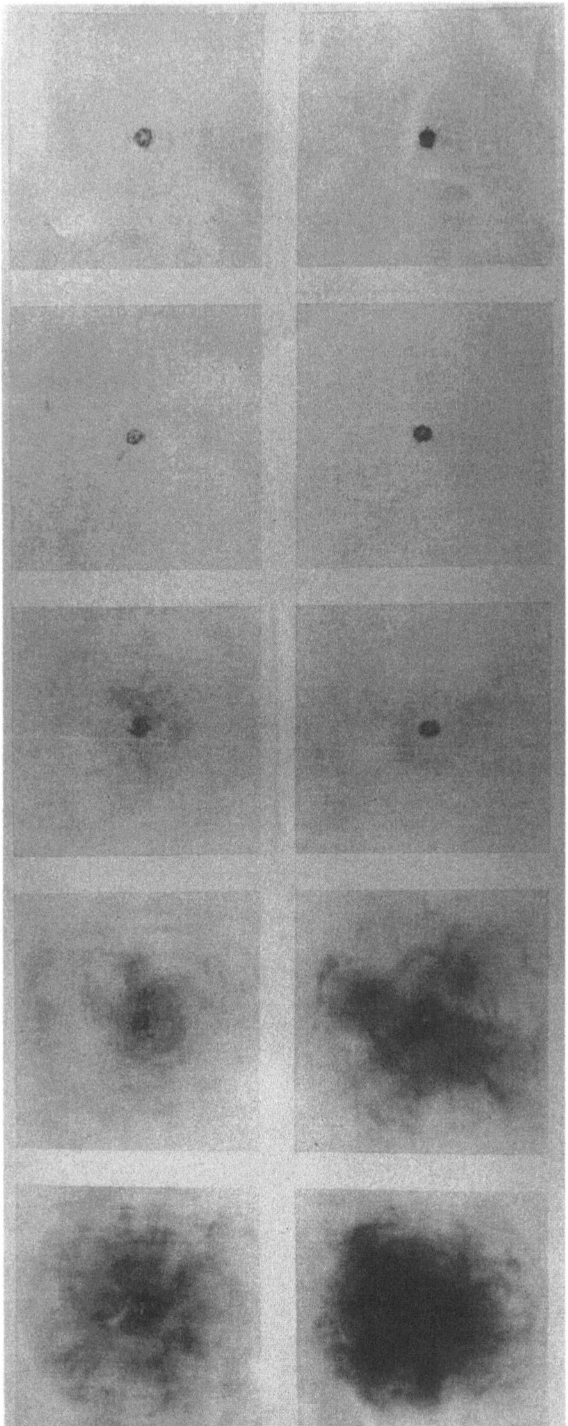

Fig. 11. Influence of barrel length on soot pattern. Smith and Wesson revolver .38 special. Left column 2″-barrel, right column 4″-barrel

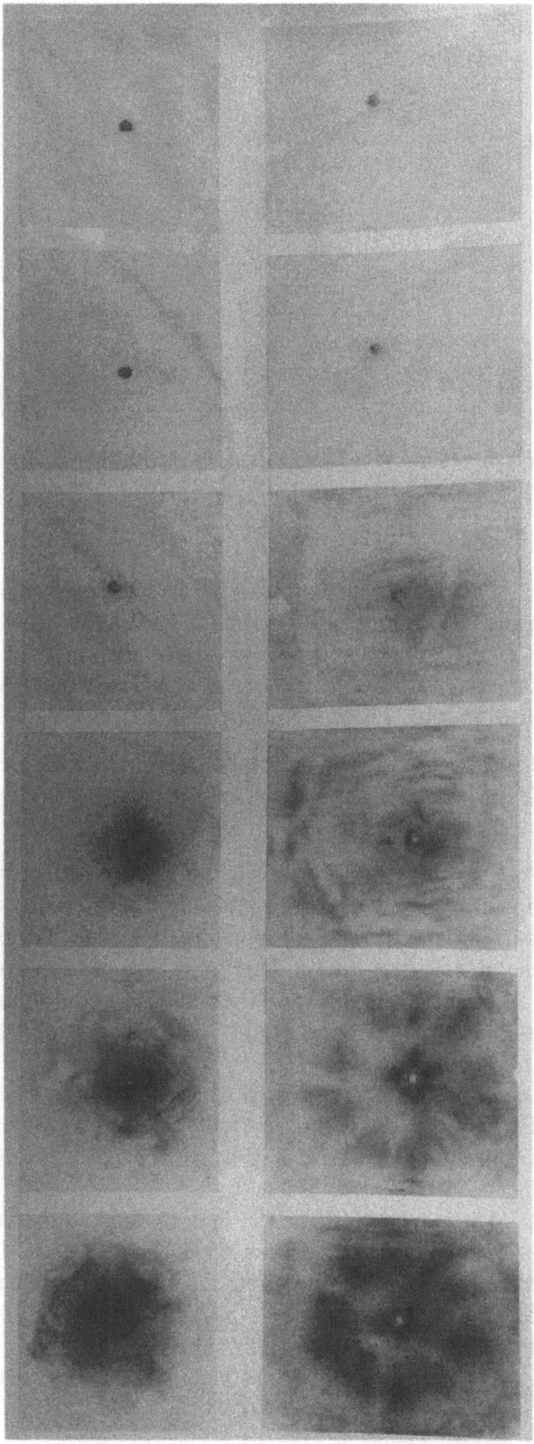

Fig. 12. Influence of various types of ammunition on the soot pattern from the same weapon. Smith and Wesson .44 Revolver. Left column .44 special cartridge, right column .44 Magnum

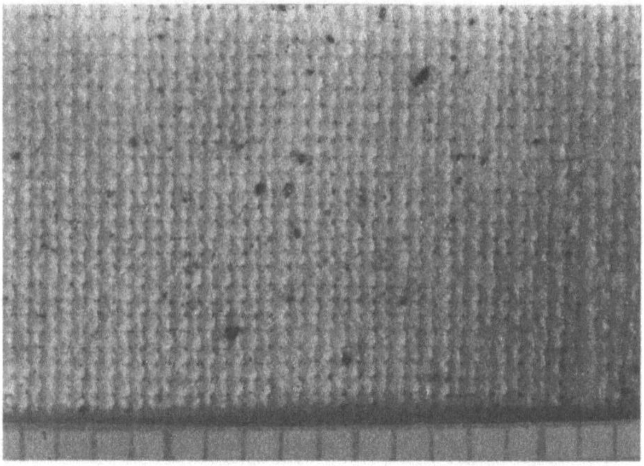

Fig. 13. Tattooing by burnt powder particles on white cloth. On the lower edge of the picture is a mm-scale

The shot range can also be determined from the hit-pattern of the powder grains according to its diameter and density on the target.

Powder soot and powder grains are not the only things to be found on the target at a suitable distance (i.e. at a close range). Metals or their compounds are also present. These metals originate from two sources:

1. From the primer ingredients (Sect. 2.1.1.1) and
2. From bullet abrasion (from the scraping of the inner surface of the barrel by the bullet)

As the bullet travels through the barrel, material is abraded from the bullet and from the barrel too, albeit in considerably smaller amounts. Because the barrel material (steel alloy) is much harder than the bullet jacket (brass, gilding metal, soft iron, nickel, lead) abrasion of the barrel is kept within limits.

All these metals—and their alloys or compounds—strike the target together with the powder-soot (and the powder grains) when distance is less than intermediate range.

Those metals (elements) which come from the primer are called soot metals. Earlier this was mainly mercury; nowadays with the SINOXID primers, lead, antimony and barium; and with the SINTOX primer, zinc and titanium (Sect. 2.1.1.1).

Those metals (elements) that come from the bullet abrasion (and the barrel) are also called soot metals but in a broader sense. They are mainly lead (pure lead bullets such as revolver and .22 bullets), copper and zinc (from brass jacketed bullets), nickel (from nickel jacketed bullets), and iron (from soft iron jacketed bullets and from barrel abrasion).

One can see that some metals can be present in both the primer and the bullet material (e.g. lead, zinc).

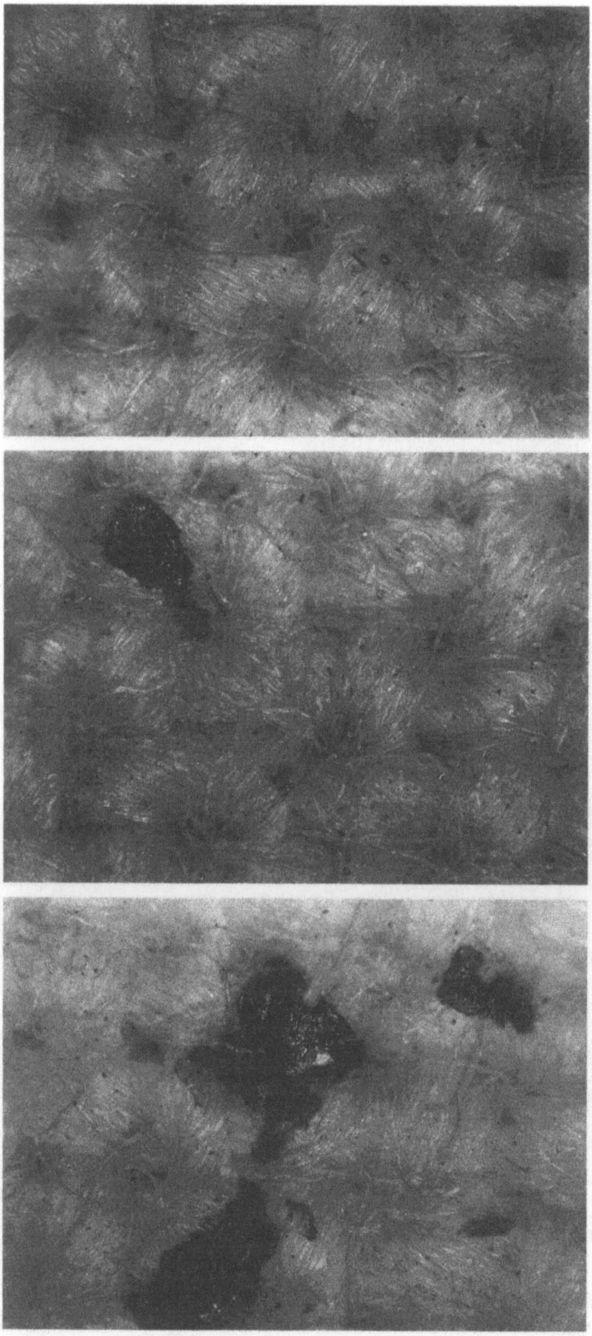

Fig. 14a–c. Powder soot blackening of cloth, highly magnified. The bigger particles are mainly burnt powder particles. The small pellets are from molten droplets of lead, antimony and barium and their alloys

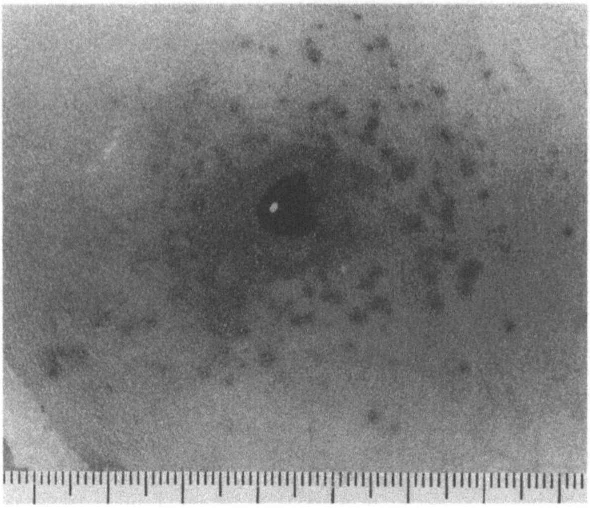

Fig. 15. The effect of powder particles on the skin. Most of the actual particles have dropped off but have left behind a dryness of the skin which has restricted blood supply and cannot be washed away. Smith and Wesson .38 special, 2"-barrel, shot range 25 cm, diagonal shot less than about 45° from "8 o'clock", there is therefore no radial symmetry of the distribution (from one of my cases)

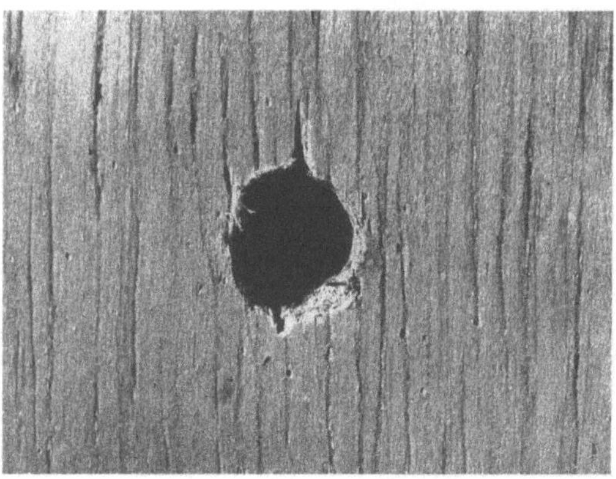

Fig. 16. Powder tattooing of a door with a soft veneer. Small calibre rifle shot, entrance hole approximately 6 mm in diameter. Shot range shown by the powder grain distribution to be about 120 cm (from one of my cases)

As a uniform contaminant of man-made fiber textiles, antimony is often present. Lead from the environment is regularly found in clothing. These metal traces which do not originate from a shot must obviously be taken into consideration during analyses.

It has just been mentioned that these metals come partly from the primer and partly from the bullet (jacket) material. It is an interesting question whether the distribution of these metals, from two different sources, is the same on the target or different. Nevertheless, the essential soot metals come from a very hot source (detonated primer) and stream with the powder gases, which are also hot of course, behind the bullet. It is therefore quite possible that, because of the differing mechanical and thermal histories, that a different metal distribution results on the target (Lichtenberg, pers. message). These considerations are (at least up to the present) of only theoretical importance.

Some authors (Nesbitt et al. 1976, Andrasko and Maehly 1977, Matricardi and Kilty 1977, Tassa et al. 1982, Wolten et al. 1979, Basu 1982) have investigated the behaviour of soot metals during a shot. The reason was not to obtain knowledge for shot range determination but rather to make it possible to discriminate shot residues from environmental pollution products by examination of the firing hand. The authors established that the elements lead, antimony and barium, partly alone partly as alloy, mainly as balls with diameters varying from < 10 to $50\,\mu m$ were present (see Fig. 14). The metals are not homogeneously deposited on the target. This has certain consequences for the accuracy of shot range determinations. At short distances, the variations in the number of these balls on the sample area is small (the variation is proportional to $1/\sqrt{n}$). At longer distances ($> \approx 1$ m), in contrast, it is large especially for small samples areas (reduction of n). The distribution can be so different that in one sample there is a ball and in another from the same shot none at all. The mass of a ball with a diameter of $10\,\mu m$ (i.e. 10^{-3} cm) is approximately 3 ng, an amount that is met with on an area of $1\ cm^2$ from a shot from a range of 1 m (Fig. 17). From such observations the variations in the measurements from shots at greater ranges can be estimated. These observations show, at the same time, that great care is essential with the interpretation of results.

The law of reduction of area density with increasing range was investigated by Baumgärtner et al. (1963) by ESA (for close range shots) and NAA (for longer ones) (Fig. 17). The weapon used was a Walther P38 pistol with a parabellum (Luger) 9 mm cartridge and annuli with an internal radius of 2.5 cm and an external radius of 5 cm measured from the center of the entry hole were investigated—a rather large area.

The two different shaped curves on the graph—from 2 to 100 cm and 100 to 300 cm (see Fig. 17) show that two different mechanisms are involved in the spreading out of powder soot (as had been previously suspected). The trend of the graph up to 100 cm derives from the fact that the soot spreads out on its own due to its high velocity till it reaches this point. At larger distances it is carried along in the reduced pressure space behind the flying bullet and therefore loses less of its intensity. The fall of the curve in this zone (above 100 cm) is not so steep. At 3 m the curve ends not because the limits of detectability have been reached, but because the amount of antimony at 3 m caused by the shot is the same as the amounts which occur in clothing material. Naturally the extension of these two zones is dependent on the weapon (barrel length) and the kind of ammunition used (Fig. 18).

Kijewski and Klewe also investigated the mechanism of soot spreading. They measured the area density of antimony with AAS (with the metal-hydride system)

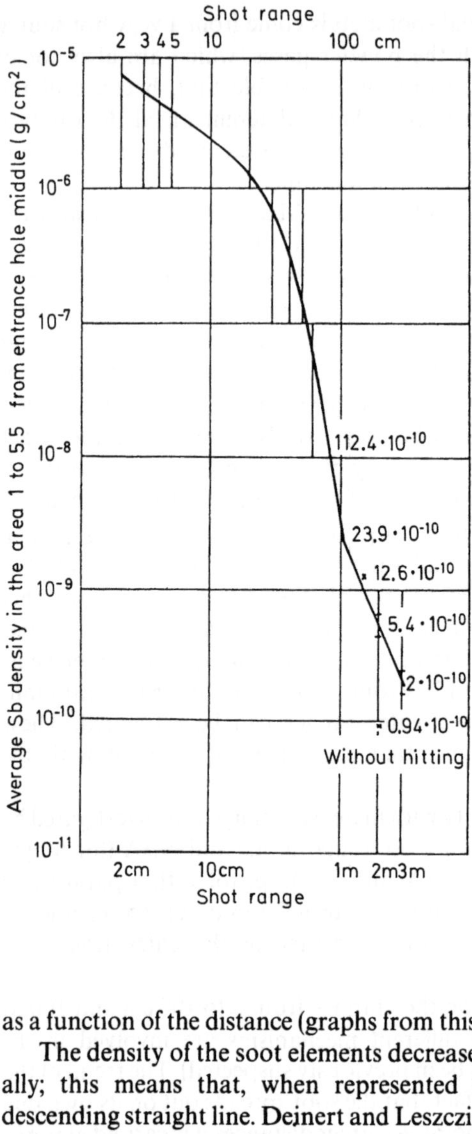

Fig. 17. Average Sb density ($g \cdot cm^{-2}$) as a function of the shot range. The weapon was a P38, 9 mm Luger, measured by ESA (for close ranges) and NAA (for long ranges). Original figure according to Baumgärtner et al. (1963). Compare with Fig. 38

as a function of the distance (graphs from this paper are shown in Figs. 18 and 38).

The density of the soot elements decreases radially almost exactly exponentially; this means that, when represented logarithmically, it is shown by a descending straight line. Deinert and Leszczinski (1980) investigated the relations with greater accuracy. They measured the extinction E of the lead pattern as a function of the shot range and of the distance r from the middle of the entrance hole at a specific shot range. They were able, with the help of their experimental results to formulate the following equation:

$$E = E_0 \cdot \exp(-B \cdot r) \qquad \text{(see Fig. 19)} \qquad (1)$$

For a specific shot range, E_0 is constant. The dependence of this value on the shot range again follows an exponential law:

$$E_0 = E_{00} \cdot \exp(-d \cdot s) \qquad (s = \text{shot range}) \qquad (2)$$

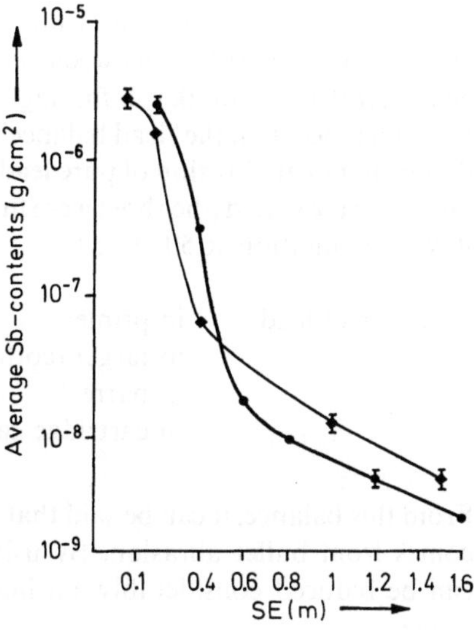

Fig. 18. Calibration curve: Sb g·cm^{-2} (y-axis) plotted against shot range. ● Colt revolver. ■ Walther pistol 7.65 mm. The curves have differing inclinations due to the different mechanisms of spreading (from Kijewski and Klewe)

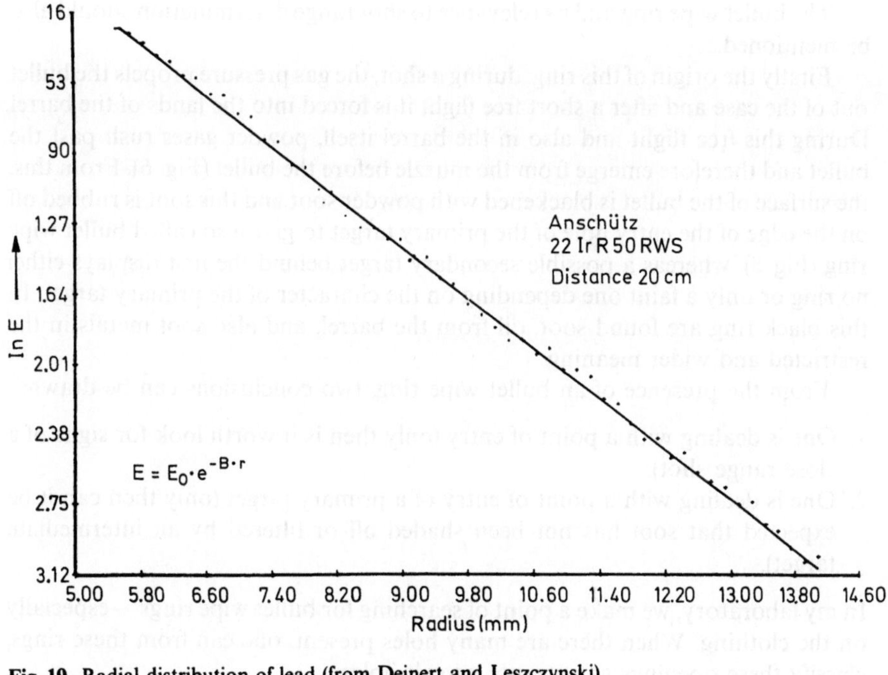

Fig. 19. Radial distribution of lead (from Deinert and Leszczynski)

This means that plotting the value E_0 as a function of the shot range gives a descending straight line (between certain distances) as common experience also demonstrates. A remarkable finding, among others, is reported in the cited work. It is concerned with the "lead balance", i.e. where can the lead from the primer and that from bullet abrasion of pure lead bullets be found after firing. The following findings were stated (the shots were fired from an Anschütz small caliber rifle with RWS ammunition R 50 .22 lr.):

Amounts of lead	in primer	$2800 \pm 700\,\mu g$
	in target (cotton)	$4700 \pm 1300\,\mu g$
	in barrel	$620 \pm 100\,\mu g$
	in cartridge case	$590 \pm 100\,\mu g$.

From this balance, it can be said that roughly 50% of the lead found in the target comes from bullet abrasion. By using jacketed bullets, the expulsion of lead can be reduced considerably, an important factors in the case of indoor rifle ranges.

In the past, authors concerned themselves with powder soot distribution and its dependence on the shot range (Baumgärtner et al. 1963, Sellier 1967, 1988, Krishnan 1967) e.g. Fig. 39.

The radial distribution of the soot metals—like that of the soot—corresponds approximately to a exponential distribution (Baumgärtner et al., 1963; Sellier 1967, 1988; Krishnan 1967; see Fig. 39).

The bullet wipe ring and its relevance to shot range determination should also be mentioned.

Firstly the origin of this ring: during a shot, the gas pressure propels the bullet out of the case and after a short free flight it is forced into the lands of the barrel. During this free flight and also in the barrel itself, powder gases rush past the bullet and therefore emerge from the muzzle before the bullet (Fig. 6). From this, the surface of the bullet is blackened with powder soot and this soot is rubbed off on the edge of the entry hole of the primary target to give a so called bullet wipe ring (Fig. 8), whereas a possible secondary target behind the first displays either no ring or only a faint one depending on the character of the primary target. In this black ring are found soot, oil from the barrel, and also soot metals in the restricted and wider meaning.

From the presence of an bullet wipe ring, two conclusions can be drawn:

1. One is dealing with a point of entry (only then is it worth look for signs of a close range shot).
2. One is dealing with a point of entry of a primary target (only then can it be expected that soot has not been shaded off or filtered by an intermediate target).

In my laboratory, we make a point of searching for bullet wipe rings—especially on the clothing. When there are many holes present, one can from these rings, classify these openings as entrance or exit holes.

2.2.1 Diagonal shots, Deviations from the Normal Pattern of Powder Soot Blackening

With the diagonal shots the radial symmetry of the soot patterns is no longer present. The soot pattern on the side further from the weapon becomes pulled out in the radial direction when the angle of the shot is more than approx. 45 °C—this follows quite clearly from geometrical consideration of Fig. 22. When the angle of the shot is less than approx. 45 °, the soot is pulled out to the side nearest the weapon (Elbel). This result ("reversal of the soot pattern") is brought about by the effect of the bullet wipe ring. The intensity of this ring on the "shooting" side is much greater due to the small angle of the shot when it strikes the target material. The bullet wipe ring contributes considerably to the powder-soot blackening on the shooting side (Figs. 21, 23).

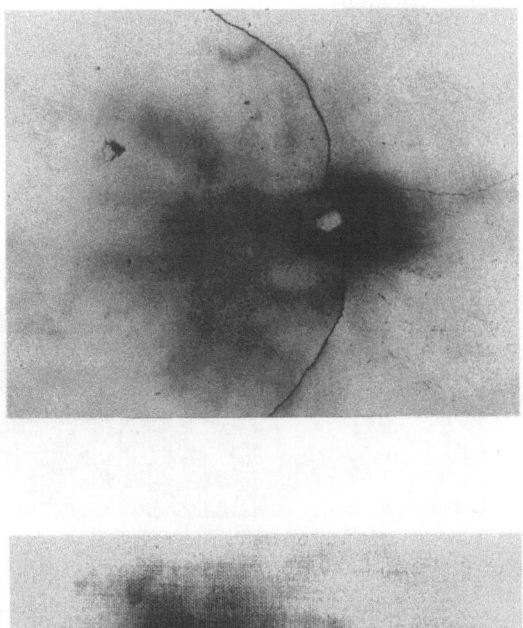

Fig. 20. Soot pattern from angled shots. The soot is spread out to the side away from the firer = the effect of the soot stream. The vortex ring is evident, however, on the side nearer to the firer. Weapon: S&W-revolver .38 spec., 4″-barrel, shot range 10 cm, firing angle 60°, firing direction from the left (Tear in the material caused by the powder gases)

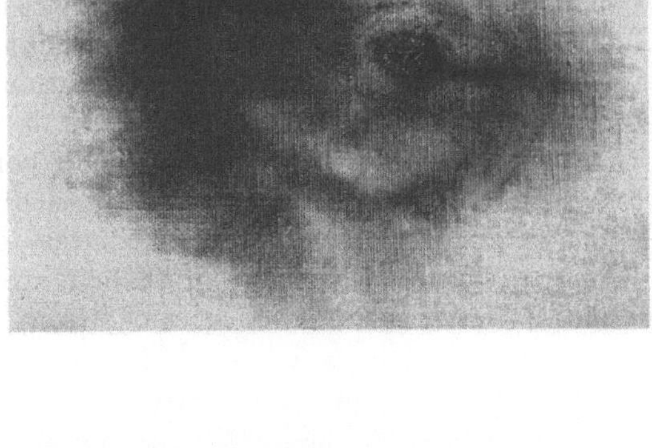

Fig. 21. Soot pattern from angled shots. The soot is spread out to the side nearer to the firer = effect of the vortex ring. Firing angle 45°, otherwise as in Fig. 20

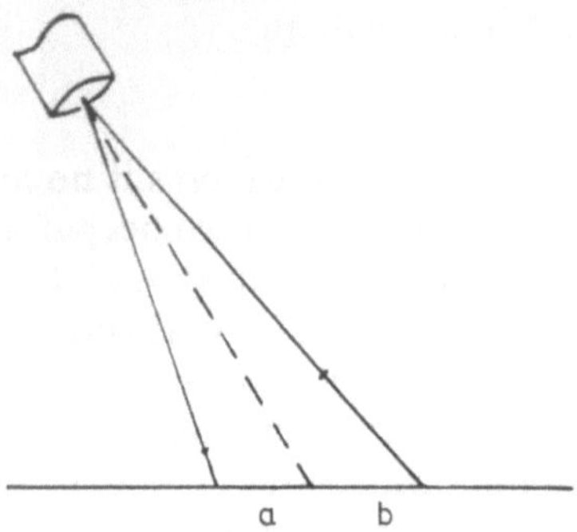

Fig. 22. Explanation of the spreading out of soot to the side nearer to the firer by large firing angles. The distance b is greater than a. This geometrical interpretation appears insufficient. With the soot stream striking at an angle a diversion to the side away from the firer takes place and leads there to greater powder soot blackening. See also Fig. 23 which clearly demonstrates this diversionary effect of the central stream

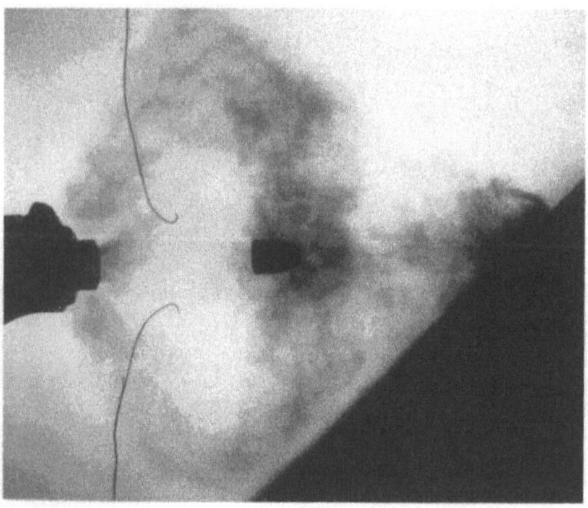

Fig. 23. Explanation of the spreading out of soot to the side nearer to the firer at small firing angles. Due to the very angled direction of the weapon in relation to the target area, the lower part of the vortex ring reaches the target considerably sooner and relatively closer to the weapon. This produces a high soot density on the side nearer to the firer. Notice, also, the powder particles at the muzzle

The pulling out of the soot pattern on the side away from the weapon at greater angles is caused by the central stream of the soot. The pulling out of the pattern to the shooting side at small angles is caused by the soot ring (Fig. 20). From this no general statement can be made about the angle at which the reversal of the soot pattern takes place with various: this is far more a factor of the intensity and distribution of the soot in the soot stream and soot ring, which, already mentioned, is dependent on the barrel length and caliber. For long-barreled weapons, the soot ring is very small and of low intensity and therefore the reversal effect is only small in contrast to the short-barreled weapons. The reversal effect is also dependent on the range, as the intensity of the soot ring is reduced faster than that of the soot stream.

3 Classification of Shot Range Zones

In practice, shot ranges are separated into 3 zones according to the changes to the target caused by either the effect of the powder gases or through the constituents and/or residues of the propellant and/or primer.

The 3 zones are:

1. contact gunshot
2. intermediate range gunshot
3. distant gunshot

3.1 Contact Gunshot (Shot with Muzzle Contact)

The changes brought about in the target by a contact gunshot are, when we ignore those produced by the bullet, practically speaking caused only by the mechanical effect of the powder gases which are still under high pressure at the muzzle. Due to the pressing of the muzzle into the target (skin, cloth covered skin), the powder gases cannot dissipate into the air.

It is advisable to consider the effect on the naked skin and the clothed skin separately because the patterns produced in both cases are different to each other, With a contact shot on the naked skin, the bursting-open effect is present in the skin; with a contact shot on the cloth-covered skin, the effect is produced in the clothing—the underlying skin is only blackened and there is no laceration.

3.1.1 Contact Shot on Naked Skin

On the human body, a contact shot—especially to the head where there are extensive areas of bone under the thin scalp—is characterized by a laceration under which lies a soot cavity and there is undermining of the skin (the skin becomes detached from the underlying tissues). Due to the sealing of the muzzle pressed against the skin, the high-pressure powder gases following the bullet force their way into the connective tissue layer between the surface of the bone and the

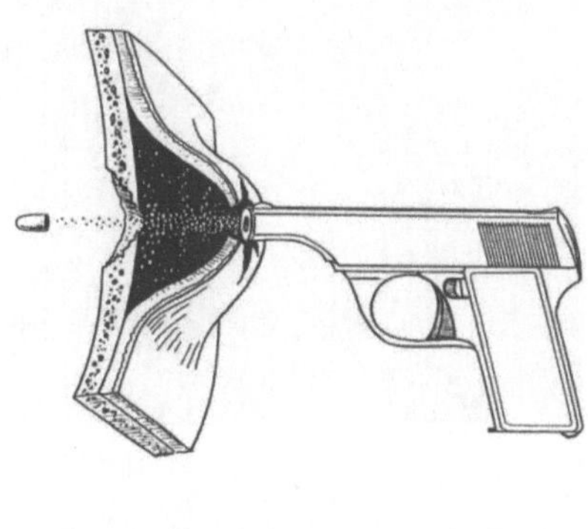

Fig. 24. Contact shot on naked skin. Characteristic signs: star-like indentation, undermined skin, soot cavity (temporary)

scalp. This raises the scalp so that it bulges towards the mouth of the gun. This section of the scalp is thereby overstreched and tears radially (Fig. 24).

This cavity effect is easy to understand when one compares semi-quantitatively the volume of the bullet's path with the volume of the following powder. The bullet's track has, for example, a length of 5 cm (when the gases appear at the muzzle) and with a 9 mm caliber a volume of about 4 ml. One gram of powder produces approx. 1000 ml gas. A 9 mm cartridge therefore produces

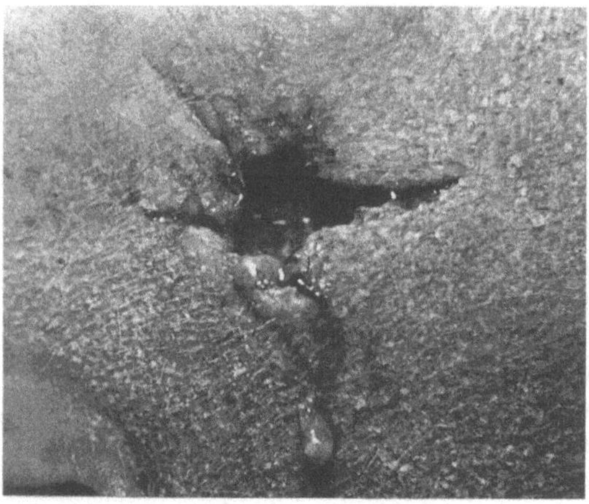

Fig. 25. Contact shot in the temple. Because the weapon was being pressed against the skin, no soot is present on the skin. Note the pulp of tissue flowing out.

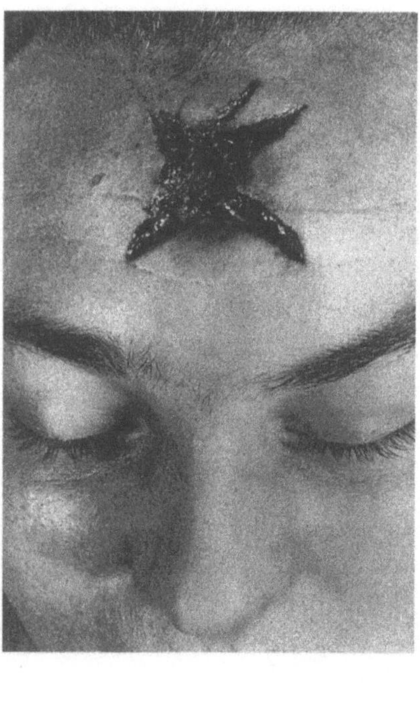

Fig. 26. Contact shot in the forehead. Typically, there is a cruciform wound

with its charge of about 0.3 g, in the order of 300 ml gas. A comparison of the volumes of the bullet's track and the gases shows that the gases create their own cavity because the volume of the bullet's track is negligible.

From the considerations, one can conclude that a contact shot has the following typical signs:

a) undermining of the skin;
b) a stellate indent wound (Figs. 25 and 26) and
c) a soot cavity with a volume proportional to the released gases.

When the powder charge of the cartridge used small—and therefore also the gas volume—the sign a) to c) may be missing. That is, for example, almost always the case with .22 short, .22 l.r., or also often with 6.35 (.25) cartridges. In contrast, these classic signs can always be observed when 9 mm Parabellum (Luger), .38 spec. (and larger e.g. .357 Magnum etc.) cartridges have been used. When these typical signs are not present, other features of the contact shot have to be searched for. Luckily, such signs do exist which show unequivocally that it was a contact shot viz. 1. the muzzle imprint and 2. powder residues present in the bullet's track.

3.1.1.1 Muzzle Imprint

The muzzle imprint is formed by the skin being forced hard against the face of the weapon by the powder gases. In this way, the epidermis at this location is injured. As a result, the skin dries out and is featured by having a brown coloration and being more leathery than the surrounding skin. Often this drying of the skin does not appear immediately after the shot, but only when the blood or tissue fluid that has issued has been removed, e.g. the wound is dry.

The muzzle imprint is, when present, a sure sign of a contact shot (Figs. 27 and 28). Weapons that have a face of the barrel which is unique can be positively identified from the shape of the dried skin (Fig. 29).

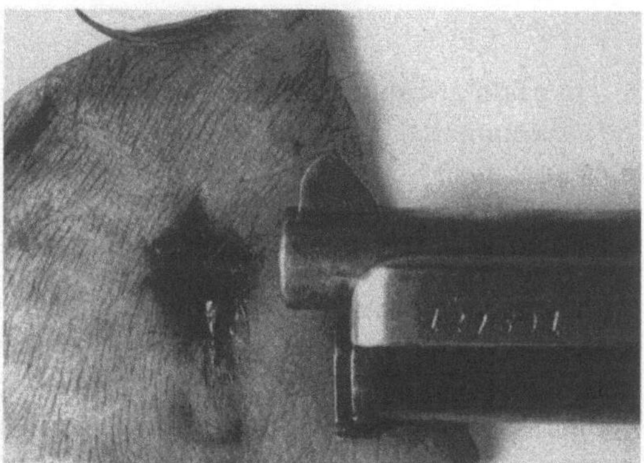

Fig. 27. Muzzel imprint. Notice the image of the front sight and the lower part of the face of the barrel

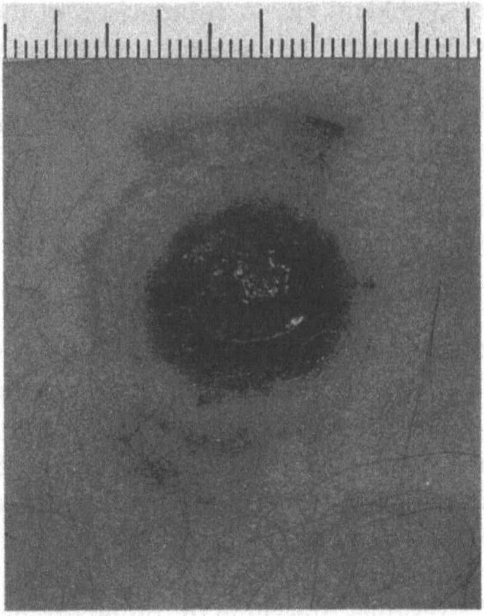

Fig. 28. Muzzle imprint from a rifle (.22 lr.). The long outer arc is caused by the muzzle and the shorter, thicker arc which lies above is from the front sight ramp

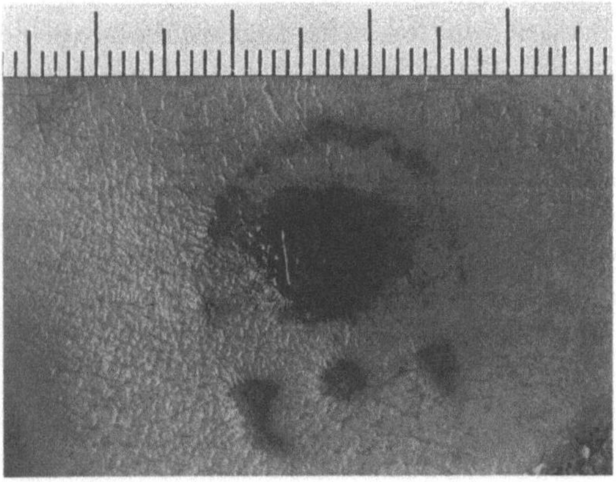

Fig. 29. Muzzle imprint, typical of a Czeska pistol, Model 27 of 7.65 caliber

In passing, it should be said that the muzzle imprint is not a vital sign i.e. it is no indication that the victim was living when shot. These can also be produced on a corpse.

3.1.1.2 Soot in the Bullet's Track

As described in Sect. 3.1.1. the powder gases from a contact shot enter the track under very high pressure. This effect is also present when—from low energy cartridges—the amount of gas is insufficient to produce a soot cavity. When a contact shot is suspected or must be ruled out and the classic evidence (indented wound, soot cavity, undermining) is missing, a sample must be taken from the wound area within the bullet's track.

It is inadvisable to introduce the cotton swab through the entrance hole as the sample could be contaminated by the bullet wipe ring.

During an autopsy, one can often see with the naked eye blackening due to powder residues in the bullet's track. This is a certain indication of a contact shot.

When making a chemical or physical examination for soot metals, the following must be taken into account. Lead in the bullet's track is not proof of the presence of soot when the bullet was of lead—the lead (and possibly antimony) could have come from the bullet itself. Proof in such cases is only given by the presence of a "non-bullet element", e.g. barium from a SINOXID primer. One should also consider that the soot elements (as evidence of a contact shot) do not come from the powder residues deposited on the bullet as these have already abraded away by the skin and do not enter the bullet's track. Thus, soot elements found in the bullet's track are not from the bullet but are transported there by the powder gases.

3.1.1.3 Powder Particles in the Entrance Hole and the Bullet's Track

At close range (< 10 to 20 cm) the powder particles and their residues are found not only on the skin surrounding the entrance hole but also in the bullet's track.

Therefore the number of powder particles varies according to the shot range. Experiments carried out by Barz (1971) on this phenomenon showed that the distribution of these particles was apparently paradoxical i.e. the particle density was low at first and increased as the shot range become greater. These missing particles had disappeared into the wound opening.

The greatest range at which such particles are still found in the bullet's track depends on the barrel length (bundling of the particles), the ammunition used, and the strength of the primer in concrete cases (the weaker the primer the more incomplete is the combustion). I do not know of any values or investigations that have been reported in the literature. However, what can be said is that when powder particles are found in the bullet's track, we are dealing with a nearer intermediate shot. The powder soot blackening, powder particles and soot elements are always found on the target at this range makes the confirmation of powder particles in the track of little interest except when the clothed skin has been shot at and the clothing is no longer available.

3.1.1.4 Remarks on the Ranges at Which the Term Contact Shot Is Valid

It is not only with the weapon pressed hard into the target that the pressure of powder gases produces a soot cavity and soot is present in the bullet's track. This is also the case when there are a few millimeters between the surface of the skin and the face of a barrel. The force of the emerging gases is still great enough to produce the same effect as the gun being pressed hard into the skin. Exact values of the ranges when these effects, e.g. indented wound, are produced cannot be given as they are dependent on the weapon and ammunition used.

How hard the muzzle was pressed into the target can be estimated from the powder soot blackening of the skin around the entrance hole. A hard-pressed weapon prevents it taking place; when the muzzle is just touching produces extensive blackening around the entrance hole. The intensity and sizes vary in individual cases and then test shots are necessary for exact statements.

3.2 Intermediate Range Gunshot

The intermediate range gunshot is characterized by the presence of powder residues and soot elements in the target and around the entrance hole. These together as a deposit in the target are known as intermediate range indicators.

The bullet wipe ring is not evidence of an intermediate range shot—it is brought to the target by the bullet and not by the powder gases. Even at ranges when no powder residues reach the target, e.g. 10 or 100 m, there is a bullet wipe ring present on the edge of the entrance hole (i.e. when the bullet has not already passed through another target).

Whereas, with a contact shot only the presence of a phenomenon (e.g. soot cavity, soot in the track of the shot) has to be established (qualitative testing), with an intermediate range shot the amount of substance (powder residues) has to be determined (quantitative testing). That means that the determination of the distance of an intermediate range shot is more expensive than a contact shot because of the apparatus required. A contact shot can in favorable circumstances be determined with the naked eye.

The expression "intermediate range" needs explaining and here several definitions are possible. It can be said that the intensity of the powder soot blackening at a specific range is proportional to the barrel length, the cartridge used etc.

However, the extent of this shot range where powder residues can still be detected is also dependent on these factors.

The size of this zone, within which the range can be determined is naturally affected by the sensitivity of the detection method. When no powder soot blackening or powder tattooing can be seen, neutron activation analysis (NAA) can still detect large amounts of lead as soot elements. Visually, for example (for a specific weapon-ammunition-combination), the furthest extent of an intermediate range shot is only 30 cm, but with NAA, however, 150 cm.

In order to determine the shot range the following factors can be employed:

1. Powder soot blackening.
2. Powder particles.
3. Soot elements.

At small distances all three constituents of the powder residues are present (close intermediate range gunshot). At greater ranges, the first to disappear is the soot; powder particles, due to their higher sectional density, travel even greater distances (distant intermediate range gunshot). When no powder residues can be detected in the target, we speak of a distant shot.

In the following Chaps. 4–6, range determination by the various methods and their sensitivity will be discussed in detail.

3.3 Distant Gunshot

The distant shot is defined negatively as being the range at which the various signs indicating an intermediate range shot are no longer present.

A layman assumes that a distant shot is from a great range and is surprised to learn that in its criminal sense that this is not so; for small weapons (6.35 pistol, 2″ revolver) a distant gunshot can, according to the detection method, be as little as 50 cm.

4 Qualitative Detection of the Signs of a Close Range Shot

This section is concerned with qualitative methods of determining the signs produced by a close range shot. The exact range cannot be so determined, only the exclusion of a distant shot. The close range shot, as previously mentioned, is characterized by the presence of powder soot, powder particles, and soot elements on the target.

4.1 Detection of Powder Tattooing

The forced-in powder particles which cause this tattooing contain nitrite whether they are from modern nitro powder or from black powder. The presence of nitrite can be tested for with the diphenylamine-sulfuric acid reaction or with Lunges reagent. The methods stated here are, to some extent, only of historical interest (and to some extent, have unfortunately become sadly famous—"The paraffin test"). However, for the sake of completeness, they should be carried out. In every case, we attempt to determine the range with the quantitative methods documented later.

4.1.1 Diphenylamine-Sulfuric Acid (DS) Reaction

This reaction is used to detect nitrites, a chemical group found in both black and nitro powder.

Principle: Diphenylamine $(C_6H_5)_2NH$, is oxidized by nitrites to give a blue dye.

Reagents: Pure concentrated sulfuric acid is placed in a porcelain dish and a few drops of water are added. A few grains of diphenylamine are then dissolved in the solution (the amount is not critical).

Method: Particles from the powder tattooing are extracted with forceps and put into the dish. Blue streaks on the particles are proof that nitrite, and therefore powder, is present. A hand lens is often necessary in order to find particles (and possible identify them by their shape). Microscopically small particles can be brushed off with a (clean) toothbrush so that they fall into the dish. Pure powder soot does not give a reaction.
Possible sources of error: During the brushing, the surrounding areas must also be investigated and a sample taken from the part of the clothing (or target) that was not clearly not affected by the shot. When, on brushing the area that was shot, many streaks appear, and on other parts none or only isolated ones, this is a proof of powder particles being present and so demonstrating that an intermediate range shot has taken place. Iron gives the same positive reaction (e.g. in cigarette ash).

4.1.2 Lunges Reagent

This reagent also demonstrates the presence of nitrites but the test is more specific than the DS reaction.

Principle: Nitrites in a weak acetic acid solution together with sulfanilic acid produce a diazonium salt that binds with α-naphthylamine to give a red azo dye.

Reagents: The reagent is prepared from a two part solutions.

I: 0.5 sulfanilic acid in 150 ml 30% acetic acid,
II: 0.1 g α-naphthylamine is boiled in 20 ml distilled water to produce a solution. This solution is then decanted from the blue violet residues and diluted with 30% acetic acid.

Solutions I and II are mixed in equal parts to give *Lunges reagent*. Also necessary is a solution of 5 g KOH in 100 ml absolute alcohol.
Method: The material under investigation is placed for some hours in the alcoholic solution of potassium hydroxide in order to extract any nitrite present. The liquid extract is then mixed approximately 50/50 with Lunges reagent. When nitrite is present, the red coloration appears in 15 min at the latest. The test is very sensitive.

Possible Sources of Error: This reaction is specific for nitrites. Not just the presence of gunpowder (or rather its nitrite fraction) is proved but that of all other nitrites as well. Traces of nitric esters will also be shown e.g. nitroglycerine, ethylene glycol dinitrate, and pentaerythritol tetranitrate. Cigarette ash, because of its nitrite content also shows a positive reaction. Owing to the high sensitivity of this test, blind samples should always be taken and tested as well. The nitrite reaction using Lunges reagent is also suitable for testing pieces of skin preserved in formalin, although not with the same specificity.

4.2 Chemical Detection of the Soot Element Pb

The dithizone method is still the most common colorimetric test for lead. It is rather time consuming and costly. I refer those interested to the literature on the subject. The method is comprehensibely described in the monograph by Koch and Koch-Dedic (1964) and also in Sellier (1967/1988).

Later, Suchenwirth (1972) announced a sheet printing method (see Sect. 5.3) based on lead detection with rhodizonate. Elements which can invalidate this lead-detection method are barium, calcium, silver and tellurium. The last two mentioned are definitely not shot-specific.

5 Morphological Methods of Shot-Range Determination

5.1 General Principles

As already mentioned in Sect. 2.2, the shape, form and density of the soot pattern, powder tattooing and the soot elements, together with other factors (cartridge type, barrel length etc.), are dependent on the shot range. From these findings, the expert can, with knowledge of the weapon and ammunition, estimate the shot range with the naked eye.

The ideal case is a shot into white cloth because the soot pattern is then readily visible. This is in contrast to shots into dark or blood-soaked cloth in the vicinity of the entrance hole. In these cases the pattern is no longer visible to the naked eye. With distant intermediate range shots, when only powder particles are still present at the target, their distribution is often hardly to be seen.

However, there are various methods of making this more or less invisible pattern visible or making it stronger.

1. *Infrared photography* brings out this soot pattern from the dark background of cloth (blood does not interfere because it is transparent to IR light).

2. *Sheet printing methods* make it possible to see the soot elements and their distribution around the entrance: simultaneously, this method also shows the distribution of the powder particles, as these particles are always contaminated with soot elements.

3. *Autoradiography* confirms the presence (after applying the NAA) of soot elements (with the exception of lead), their location and their intensity i.e. their distribution.

4. *X-ray pictures.* With the help of these methods, which will be described in detail, the total pattern of the shot residue distribution (whether soot, soot elements and/or powder particles—according to the detection method) is obtained and then evaluated. The assessment and determination of the shot range is made with the help of test shots from the actual weapon and ammunition into the actual clothing involved in the case (naturally this would be an ideal case) from various ranges.

Evaluation of Shot Range Determination Methods. These remarks are not only valid for morphological methods but also the sampling tests describes in Chap. 6. The general criteria are:

1. Simplicity
2. Time required
3. Preparation: a) Time, b) Contamination
4. Personnel
5. Sensitivity of the test
6. Specificity
7. Documentation
8. Destruction of the samples or not during the test

5.2 Infrared Photography

On very dark or black textiles, the soot pattern, which consists of black soot, is not discernible in visible light due to dark background.

The variation in the reflecting powers of dyes and surfaces in the visible spectral range compared to those in infrared has the effect of producing different intensity values. In infrared light, soot remains black, whereas synthetic dyes generally reflect the infrared and appear bright. The soot pattern now stands out against the light background. (Fig. 30 a and b). There are two ways of displaying the IR picture:

1. *Photographs using IR-sensitive film.* IR-sensitive films are also sensitive to visible light (at least to the blue spectral range). This means that IR filters must be used. The transmission band of the filter must naturally be chosen to match the exposure density relationship of the film. When pictures are taken, it must be remembered that the ranges marked on the focusing ring of the lens are not correct for IR light. It must therefore be adequately stopped down. Formerly, Leica objectives (e.g. Elmar 50 mm, f 3.5) were equipped with a IR setting (so that for a distant object, normally set at ∞, the setting is at 15 m for IR).
The emulsion is characterized by specification of the wavelength of the sensitization maximum. Optimum results, in practice, have to be obtained by trial and error until the background color (textile color) appears the brightest. This is costly in time and material.
Polaroid, however, produces an IR film and this of course, saves the time normally spent waiting for developing.

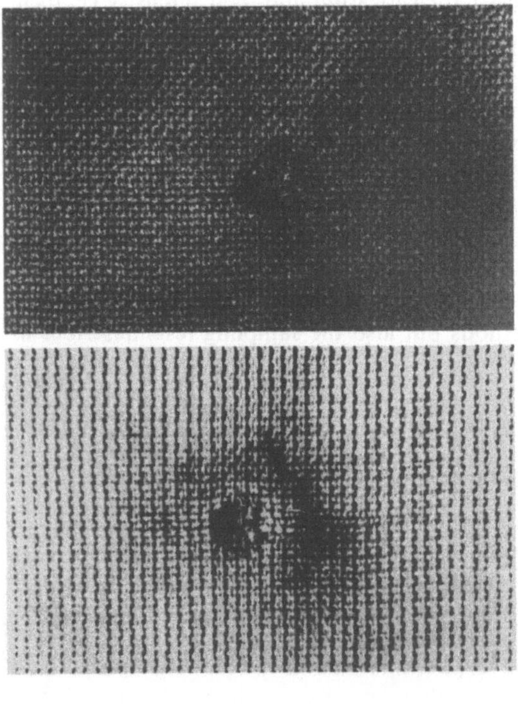

a

b

Fig. 30 a. Entrance hole in a very dark cloth. Evidence of a close range shot is not visible. **b.** An IR picture of the same cloth (at 850 nm)

2. *IR-image converter.* The IR-image is received by a IR-sensitive photo cathode, it is then electro-optically amplified and appears on the screen of the converter where it can be viewed and photographed. With the insertion of filters with various transmission bands, an optimum picture can be produced, i.e. with the highest possible contrast between background and soot pattern. IR-image converters are expensive and are only worth purchasing when they are used extensively, for this or other purposes.

5.3 Sheet Printing Methods, Chemical

The methods described in this section are used to make the distribution of the powder particles on the target visible.

5.3.1 Sheet Printing Method after Walker and also Mayer and Wölkart

Walker (1937) based his method on the nitrite reaction of the Lunge test using 2-naphthylamine-4,8-disulphonic acid (C-acid) or 1-amino-8-naphthol-3,6-di-sulphonic acid (H-acid). Normal photographic paper was then treated according to the author's instructions. When using this method, which had been developed in the USA, Mayer and Wölkart (1955) came across difficulties due to differences in the photographic paper produced in Germany. When using German paper for the Walker test, a dark red brown discoloration appeared.

Mayer and Wölkart therefore based their method on the Griess reaction with α-naphthylamine/sulfanilic acid, i.e. on the previously described Lunge reagent.
Method:

1. Normal, white, glossy photographic paper is fixed, washed well, and dried.
2. This paper is left for 10 min in a solution of 0.5% sulfanilic acid in dilute acetic acid, dabbed dry with clean filter paper, and placed in a drying cabinet at a temperature not exceeding 80 °C until perfectly dry.
3. The paper is drawn briefly through a solution of 0.5% naphthylamine in dilute acetic acid (i.e. until the paper is uniformly moist), dabbed and dried at 80 °C for a short time.
4. Finally, the paper is hardened in absolute methanol and again dried at 80 °C.

The object under examination is laid on the prepared paper and dabbed with a dry cloth. It is then ironed for 10 min, care being taken that the iron is not too hot. The presence of powder particles is shown at the appropriate places by distinctive red stains representing the surface distribution.
To achieve good reproducibility, the iron should be kept at a constant temperature. However the temperature should not be too high because when 200 °C is exceeded there is a danger that the paper will become uniformly and intensively red, so ruining the test. The prepared papers can be stored for several weeks and the necessary reagents keep for months in dark, well-sealed, glass bottles.

5.3.2 Sheet Printing Method after Leszczinski

This method allows the surface distribution of the soot elements Pb and Ba (not Sb) and some others which come from the jacket of the bullet (*soot elements in a broader sense*), e.g. Cu and Ni, to be detected and recorded.

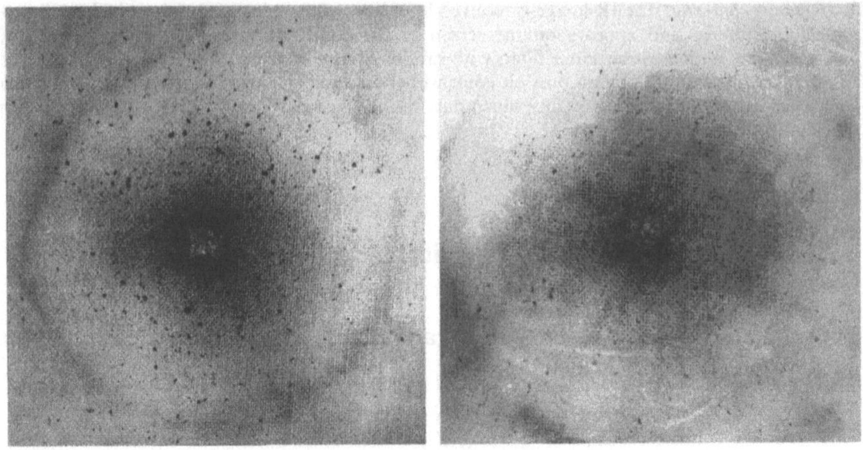

Fig. 31 a, b. Sheet printing method after Leszczynski. (**a**) Original, (**b**) lead picture

This method was developed and applied by Leszczynski (1959) in BKA (Bundeskriminalamt i.e. the Central Bureau of Criminal Investigation in Germany). Fundamental improvements were later made by Hoffmann and Lichtenberg (also BKA). (Further papers are published by Hoffmann (1985) and Lichtenberg (1983, 1985, 1988)).

In principle the method is as follows. A thin plastic foil (Cellophane P600 from Kalle and Co. Wiesbaden, 60 μm thick) is soaked for several days in a 15% acetic acid solution. The moist foil, after being dabbed gently, is laid on the trace carrier—usually clothing. A neutral plastic foil is then placed on top and this sandwich of trace carrier, prepared foil, neutral foil is placed between two felt pads and then put in a press and subjected to high pressure (as a rule, on the area of $200 \times 200 \text{mm}^2$, a force of 160 kN). The felt pads are to even out any irregularity in the surface of the sample. Due to the intimate contact the metals diffuse into the acetic acid-soaked foil and can be chemically detected there.

To develop the lead pattern, the foil is placed for 5 min in a 25% Na_2S bath (Merck-Nr.6638). In the foil can be seen brown to black (according to the concentration of Pb) lead sulfide pattern which shows qualitatively and (relatively) quantitatively the distribution on lead on the target. After being chemically treated, the foil is irrigated until the smell of the H_2S is no longer discernible—approx. 1/2 h under running water (Fig. 31 a and b).

In order to reveal the barium distribution either a 25% Na- or K-dichromate solution is used. Deep yellow Ba-dichromate is formed (N.B. when lead is present also Pb-dichromate of the same color). The foil is then irrigated until the water is colorless.

In the same manner the copper or nickel patterns can be developed (see Table 7).

The foils are squeezed on the high-gloss supporting foil of a photodrier and dried at a temperature not exceeding 80 °C (at higher temperature there is a danger of cracking).

The sheet printing method does not function when the weave of the cloth in the sample area, i.e. around the entrance hole, is full of blood, and this is often the case. When this occurs, no soot elements are detectable on the corresponding location on the film. This has led to errors in shot range determination or even to entrance and exit holes being confused. When the entrance hole itself is heavily bloodstained, metals from bullet wipe ring can hardly be detected or even not at all. An inexperienced investigator could then diagnose “exit hole” because of the

Table 7. Dyeing of soot elements by the film sheet printing method according to the developer bath

Element	Na sulfide bath	K-dichromate bath
Lead	brown	yellow
Barium	—	yellow
Nickel	grey	—
Copper	green	—

After Hoffmann 1985, color pictures see Sellier (1982)

missing bullet wipe ring. This possible error can not be blamed on the method, but rather on the ignorance of the person using it.

To prevent this error from occurring, the blood must be dissolved from the cloth by the following method (Hoffmann, 1985): The stain is sprayed with 3 to 10% H_2O_2 solution in 15% acetic acid. In this way the dried out blood can be dissolved. Afterwards, two prints should be made because, from experience, the second is better than the first.

As already mentioned at the end of Sect. 2.1.1.1. (Primer) the Pb-Sb-Ba-primers (e.g. SINOXID) have been replaced. This is due to environment reasons (e.g. shooting in indoor rifle ranges) and leadfree primers containing Zn and Ti are now used (SINTOX or others, which are similar, in the USA). If a target which has been hit by SINTOX ammunition is examined by the printing method the results are negative. Any lead detected comes from the lead core of the bullet which is not fully jacketed. Lichtenberg (1983) reported a method by which Zn could be detected by the printing method. The chemistry of the reagent used for detection (Chlorindazon DS) was first described by Schweppe. With this reagent, Cu, Ni and Zn (also Co and Cd- but this is not relevant here) can be detected.

The previously described printing method for the elements Pb, Ba, Ni and Cu must be modified somewhat for the detection of Zn because the metal complex can only form under basic conditions. Instead of acetic acid, dilute ammonia solution (pH 12) is used to swell the film.

The sensitivity is not as high as the Zn detection by the AAS method; at ranges greater than 50 cm no Zn picture appeared on the print (SIG-Sauer pistol with a 9 mm Luger SINTOX cartridge).

According to Schweppe the solution is prepared as follows:

100 mg Chlorindazon DS (CDS) is dissolved in 5 ml NH_3-solution and 95 ml water is added. The solution may be kept for several weeks and can be used for the detection of Co, Ni, Cu, Zn and Cd in dilutions up to $1:10^6$. If these cations are in still lower concentrations, the reagent is used after its dilution of 1:10 with water. The solutions of the reagent have a red-brown color. Cu gives a blue coloration, Ni and Zn give a violet one.

Lichtenberg (1983) made a remarkable observation on the Zn-distribution. From SINOXID ammunition, the lead which was shown came from the primer and from the bullet (from the open base in full-jacketed bullets and from the free-lying part of the core in partially-jacketed bullets). The portion from the primer cannot be separated from that from the bullet (this would be possible if differing Pb isotopes were used in primer and bullet core). With the use of a (lead-free) Zn-primer, the both metal distributions are easily separated. The soot-element distribution on the target consists of Zn, the bullet-element distribution consists of Pb (the Pb-abrasion from the bullet is partly mechanical due to the friction and partly thermal caused by vaporization of the material from the bullet base. The author found that the soot elements were in the central part of the powder soot pattern and the elements emanating from the bullet were preponderantly in the outer section. At present, this may only be of theoretical interest but could be of practical use in the future.

5.3.3 Sheet Printing Method After Suchenwirth (1972)

Filter paper is moistened with an aqueous solution of tartaric acid and pressed onto the trace carrier. This print is sprayed with aqueous Na-rhodizonate solution. Lead, when present, gives an intensive red coloration.

Reagents:	Tartaric acid, aqueous solution, approx. 1%, Na-rhodizonate, a saturated aqueous solution (always freshly prepared).
Detection limit:	approx. 1 in 10^6.

In the literature, Ba, Ag and Te are said to negatively influence the test. However, in my investigations, these problems have never arisen (Ag and Te would be unexpected). With white or light colored cloth it is also possible to spray tartaric acid directly onto the cloth and, after letting it take effect, proceed with the test as previously described.

Small amounts of blood can be decolorized with hydrogen peroxide solution. The sample must, however, be completely dried to destroy the H_2O_2 solution before proceeding to make the lead visible.

As the method is very simple and no apparatus is required, it can be used at the scene of crime (e.g. to test for bullet effect on walls and windowpanes).

According to the author the detection limit for an intermediate range shot for a Brown. 6.35 pistol is approx. 60 cm, for 9 mm Parabellum pistols between 80 and 120 cm.

Naturally, it must be checked to see if the detected lead is really an effect of a shot (powder soot blackening, bullet abrasion) and not from the environment and not there just by chance. Testing with sodium rhodizonate does not affect a later examination by other methods, as the lead in the form of the rhodizonate remains in place.

5.4 Sheet Printing Method, Physical (Autoradiography)

The NAA method is described in detail in Sect. 6.4. Only the essential details are dealt with here.

The principle of the method is to activate the soot elements with NAA so that a autoradiograph can be made. Unfortunately, NAA activates not just elements

Sb and Ba (Pb cannot be detected for physical reasons) but also many other elements on the target (matrix effect). These elements include, among others, S, P, Na and these can interfere with the test and their influence must be eliminated by radioanalytical or chemical methods, or at least minimized.

Naturally, the individual elements can be recognized by their characteristic β- or γ-radiation. The photo-emulsion, however, indiscriminately registers all forms of radiation by blackening whether they emanate from the elements, Sb and or Ba, which we wish to detect, or from these disruptive elements. This results in a background in which the shot pattern can be completely submerged.

Contaminants which can be involved are: Sb in man made fibers; Na and S in blood; S and P from detergents and textile finishing materials. These activated contaminants must therefore be eliminated before the autoradiogrphy.

The following method, proposed by Schmitz (1973), can be used to achieve this by transferring 50% of the traces from the shot onto an inert carrier. An adhesive film is placed on the trace carrier and is places in a hydraulic press for 1 min at a pressure of 0.5 kN/cm². The film is the removed and the sticky side covered with a 50 μm thick polyethylene film and then a autoradiograph made. The adhesive film "Mypolam Buchfolie" has given particularly good results.

With this printing method, essentially what happens is that the soot element balls (see Sect. 2.2) are removed from the trace carrier whereas the Sb present in the man-made fibers and not originating in the soot is not transferred to the film.

The same author also prescribed the following which produces a favorable density/background relationship when making autoradiographs:

Activation:	integral neutron flow—approx. 10^{17} cm^{-2}
Decay time:	approx. 10 d
Film material:	Curix RP1 clear base from Agfa-Gevaert or X-ray screen film clear base from Dumont.
Exposure time:	approx. 10 h
Printing pressure:	approx. 0.25 N/cm²
Developer:	Tetenal Eukobrom paper developer at a max. temperature of 20 °C for about 5 min.

The activation of the samples was done in the FR2-Reactor in Karlsruhe with a neutron flow of $22 \cdot 10^{11}$ cm^{-2} s^{-1} and a radiation time of 7 *d*. Due to the Na content of the samples the measurements were made after a 10 d waiting period. The amount of ^{124}Sb with its relatively long half-life of 60 d is mainly measured (see Table 10 in Sect. 6.4.1). Because of the long half-life, a number of exposure-time and measurement series can be carried out with a particular sample. There is so much more time for experiments than when the short lived isotope ^{122}Sb is used with its half-life of only 2.8 d (admittedly, the optimum radiation time would be shorter, Sect. 6.2.3).

Publications on autoradiography in shooting cases are rare. Obviously this susceptibility to failure due to the matrix influence is so strong that the disruptive contaminants in some cases cannot be estimated as shown by experience with other methods. Previously published experiments have all been carried out under favorable conditions. The above-mentioned author thus chose, in his preliminary tests, the filter paper with the smallest proportion of disruptive elements from among more than 15 others. In practice, as far as I know, this method has not lived up to expectations. Further experience with the method has not been published. A great disadvantage is also the long period of waiting for the results (7 to 10 days) plus the fact that the sample must be given to another laboratory.

The criticism does not apply to the method described later (Sect. 6.2.3) in which the soot elements after treatment with NAA can be detected by their β- or

γ-lines with very high sensitivity. The advantage of being able to show the overall distribution of the soot elements is, however, lost.

5.5 Determination from Powder Tattooing

This method comes into its own when visible powder soot blackening is not present i.e. with a long, intermediate-range gunshot. The density (i.e. number of particles/cm^2) and the distribution is determined and then compared with test shots at various ranges.

The method has some important shortcomings. The number of powder particles emitted from the muzzle is relatively small. These particles spread themselves over a relative large area so that the density determined varies randomly. Normally 5–15% of the powder emerges from the muzzle in the form of unburned or partially burned particles. This amount depends—other conditions being equal—on the structure of the primer and its ignition in the actual case.

An advantage of this method is that the powder particles or their outward signs on the skin cannot be washed away. This is in contrast to powder soot and the soot elements. This is of great importance when because of the surgical operation or due to ignorance the entrance hole and its surrounding area are cleaned.

It is interesting that the number of particles does not become steadily lower but starts at zero for a contact shot, increases to a maximum, and only then begins to decrease. The position of the maximum depends on the weapon (barrel length) and the ammunition. This effect is easy to explain—with a contact shot, all the powder particles go through the entrance wound into the bullet's track. With very close range shots (a few cm) the powder particles are still so closely bunched together that most of them enter the wound. Not until longer ranges (from between 5 to 7 cm upwards) is the spread so great the number of particles entering the entrance wound can be ignored. Barz (1971) and Barnes and Helson (1974) carried out long series of experiments and the results of the two authors are compiled in Table 8. The distance traveled by the powder particles is dependent

Table 8. Total number of powder particles as a function of the shot range for various weapons (Barnes and Helson)

	Shot range in cm (converted from inch and rounded off)					
Ammunition	0	7.5	18	33	53	79
.22 short	0	62	60	60	35	2
.22 l.r.	3	135	124	132	120	43
.25 ACP	0	25	28	15	0	—
9 mm Luger	0	95	130	27	0	—
.38 spec.	0	12	58	62	50	10
.45 ACP	0	68	52	37	18	—

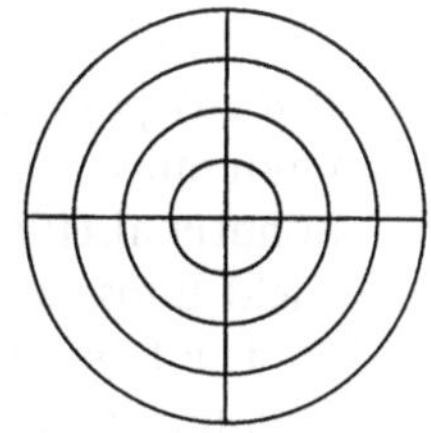

Fig. 32. Transparent plastic disc with cross-wires and concentric circles 1 cm apart for assessing powder particle distribution

on their shape. Spherical particles, because of their better ballistic shape, travel further than leaf-shaped powder (Dimaio (Gunshot wounds, Elsevier 1985)).

Barz carried out his experiments with a .22 rifle. He also found that the total number of powder particles reached a maximum at a certain distance (greater than zero). Under the conditions of his experiments this range was 40 cm. He also found that the radial distribution of the particles with increasing range experienced a characteristic change: at closer ranges the particles lie in close proximity to the entrance hole, with increasing range a large proportion of the particles are found at some distance from the hole.

The shot range determination is made from the morphological picture, especially the diameter of the hit pattern not, however, from the number of particles because within a wide extent of shot ranges the number of particles remains constant (see Table 8).

For describing and assessing the distribution, I use a simple device, as shown in Fig. 32. It can be easily made using a ruler and compasses and scratching the figure onto a disk of clear plastic—the concentric circles are 1 cm apart. The device is simply laid on the object of study.

The nature of the target has a decisive influence on whether the hit pattern is complete. Loosely woven cloth traps all the particles which strike it, a closer weave lets large particles fall especially when it is put under mechanical stress during or after being impounded. From hard, smooth, and clean materials (glass or metal), almost all the particles fall off or are removed by unintentional rubbing.

The skin is rather smooth and hard and the particles almost always fall off. They leave behind, however, a mark to show where they hit. This consists a fine circular subcutaneous bleeding of the epidermis and drying of the skin which cannot be removed even with vigorous washing. The pattern of the powder distribution is therefore firmly fixed in the skin. The area of this skin change is larger than one on the area of a powder particle and thus the pattern of the same particle distribution on skin seems denser then one on white paper, for example.

5.6 Imaging with X-Ray, X-Ray Fluorescence

In Sect. 2.2 it was said that the soot elements are in the form of small and extremely small spheres. It thus seemed reasonable to try to take X-ray pictures of these spheres in order to see the distribution of the soot. Because of the smallness of the spheres a very high definition photograph is needed. That necessitates a small focus point and film material with high resolving power (e.g. film for mammograms).

Böhm et al. (1969) investigated this problem. They used an X-ray camera from the Field Emission Company, the distance was 55 cm, exposure time 3 min at 25 kV. The film was Kodak SR 54, development time 5 min. The cloth under investigation was placed on a film which was coated on one side, with the suspected area facing the X-ray tube and then an exposure made. With this procedure one gets a picture of the distribution of the (bigger) soot element particles and bullet abrasion particles. One can so determine the shot distance with the aid of calibration pictures.

A further X-ray method for determining shot ranges is X-ray fluorescence. It is based on the fact that the inner electrons in the so-called K- and L-shells of the atom undergo excitation from the X-rays. This stimulated radiation can be quantitatively measured by emission spectrography. The sample from the target is taken but not with the bullet wipe ring because of the large proportion of soot elements make the (distance dependent) powder soot blackening inaccurate and insensitive. The limit of detection with X-ray fluorescence is, in any case, much lower then the other methods (ESA, AAS, NAA).

The upper limit of detection for 9 mm Lugar-ammunition for all the soot elements (Pb, Sb, Ba, Zn, Cu) is only approx. 40 cm. The advantage of this method is that the sample is not destroyed and the method can be repeated as often as you wish. All the elements are measured at one go, as is the case with ESA.

6 The Sampling Test Method for the Quantitative Determination of Shot Range

6.1 Initial Remarks and Underlying Principles

The procedures described in Chapter 5 involve the use of the overall pattern, be it directly after "development" in order to determine the shot range. With the sampling method, however, only a particular area (sector) of the distribution pattern is taken and quantitatively examined for soot elements.

A prerequisite for the quantitative determination of soot elements is a calibration curve obtained from shots from known ranges. These shots must be made with the actual weapon and ammunition because the condition of the barrel and the type, amount, and composition of the primer influence the results. The method is simple in principle: the area density ($\frac{\text{mass}}{\text{area}}$) of the soot elements (Pb, Sb, Ba, Cu, Zn, Ti) is proportional to the shot range and to the position of where the sample is taken in relation to the center of the entrance hole. When the sample position is kept constant then the area density is then basically only a function of the shot range.

Essential parameters are:

1. Sample taking position.
2. Sample (or examination) area.
3. Detection method.
4. Choice of soot elements.

Parameter 1.
Due to the distribution of the soot elements being an exponential one (see Sect. 2.2), the best position to investigate is as near as possible to the edge of the entrance hole as this gives the greatest sensitivity. However, care must be taken not to include part of the bullet wipe ring as this would invalidate the test as the amount of soot elements is not proportional (or minimally) to the shot range. Baumgärtner et al. (1963) demonstrated that the Sb density does not fall off immediately as in exponential distribution but rather not until 1.5–2.0 cm from the center of the entrance hole. This is apparently the best position from which to take the sample: on one hand it is far enough away from the bullet wipe ring and on the other the density is still near optimum.

Parameter 2.
For the reason of sensitivity, it is recommended that the largest area possible be taken for the use in the investigation. However, this has limits. If the area is very large and the material is destroyed during the investigation, the process cannot be repeated because of lack of material. One most always bear in mind that an investigation can go wrong and that large quantities require longer to process (incineration etc.) the sample for investigation. Therefore a compromise must be made. When judging a particular method with reference to the maximum determinable soot elements (i.e. the detection sensitivity) one must also consider the size of the sample area. It is simple to increase the sensitivity of a relative insensitive method by increasing the sample area.

So that specimens are damaged as little as possible, the area of the sample should be as small as possible. This is possible when the method used is sensitive enough. One should, however, always remember that the soot elements are not homogeneously distributed over the target (see Sect. 2.2). Thus, a too small a sample area leads to more or less large statistical fluctuations of the measurements.

With consideration of the shape of the area: since the distribution (when the shot is vertical to the surface) is rotationally symmetrical it is logical to take a rotationally symmetrical sample area, i.e. a torus or a torus sector. With the small sectors a rectangular shape can be used which is more convenient.

For taking samples from textiles, hole punchers have proved themselves as being highly convenient. The samples are simply punched out with the device and are identical in size.

Parameter 3.
All detection methods that are sensitive enough can be used when they fulfill the criteria mentioned in Sect. 5.1.

Parameter 4.
Selection of the soot elements for the shot range determination.
In currently and normally used ammunition the following elements are present: Pb, Sb and Ba, although Sb and/or Ba can be missing (see Sect. 2.1.1.1). In addition the soot elements in the wider sense which come from the bullet surface: Cu and Zn (from brass) or also Ni.

Lead is the element most commonly found in the primer and its analysis gives the greatest sensitivity for determining the shot range (i.e. with the same detection sensitivity of the method of measurement for all the elements mentioned—this is not valid for the NAA method for example, then lead is insensitive, in contrast to Sb and Ba). However, lead has a decisive disadvantage: it is ubiquitous which means that it can be found everywhere and is not specific to the primer. It can be found almost always in textiles—even those not involved in a shooting—and also on the hand which was not holding the gun. If the density of lead over the whole piece of clothing were uniform this would not affect the shot range determination. It would then be a constant factor in the calibration curve which would be displaced in the direction of the positive y-axis (i.e. area density). Unfortunately it cannot be assumed that this "background" density of lead is constant (e.g. clothing of painters—contaminated by pigments contain lead, clothing of people working in gas stations—contaminated by leaded fuel). The determination of the shot range by the lead density must be made with great care. Antimony has proved itself to be the best element because it is seldom found naturally in the environment and does not lead to contamination of the areas under investigation. Antimony is however, almost always present in synthetic fibers as it is used as a catalyst in the manufacturing process. The area density of Sb is, in my experience, constant and can be taken into consideration when the calibration curve is made. For that reason, control samples must always be taken. The control samples must be taken from locations which are at an appreciable distance from the entrance hole. Also barium, copper, zinc and/or nickel (the latter from the jacketed bullets) can also be used. Significant contamination from the environment is not to be expected.

SINTOX ammunition contains Zn and Ti. Zinc is a widespread trace element which make its presence noticeable in the AAS test. It is by this test that Zn can be detected with extreme sensitivity. Titanium is widely distributed in small quantities to such an extent that the soil in fields contains on average more than 0.5%. This must be considered when examining dirty clothing. Titanium, like iron, is a "multiple line" element. It emits and absorbs a large number of spectral lines with the result that the energy (in contrast to the alkali metals) is distributed over many lines—the detection sensitivity by emission and absorption is therefore relatively low.

6.2 Emission Spectrum Analysis (ESA)

6.2.1 General Comments

This method was originally developed to detect Pb and Sb, and to a limited extent Ba. Because of its specificity only Sb was determined.

A characteristic of the emission spectrographic method (also the AAS method) is that the density of the spectral lines on the photographic plate does not give an absolute measurement of the element present. This is due to the fact that the density is not only proportional to the amount but to several other factors as well:

1. Photographic plate characteristics
   Gradation
   Sensitivity
2. Developer characteristics
   Type of developer
   Temperature
   Development duration
3. Excitation arc
   Shape of the electrodes
   Distance between the electrodes
   Duration
   Voltage and current
4. Matrix effect (influence on the excitation by foreign matter in the sample).

To eliminate this multitude of extraneous factors a reference element must be used (internal or external standard): the spectral lines of the reference element are under the same influences as the element being investigated. In ideal cases the influences on the lines of the two elements compensate each other, so that the densities S of these lines or the quotient of the intensities I (due to $S = \gamma \cdot \log I$, γ = gradation) become independent of these influences. One must, however, make sure that the area density of the reference element is identical on the locations from where samples are taken. This requirement is fulfilled by iron, for example, as it is present in all natural fibers.

6.2.2 Choice of Lines for Sb and Fe

The Sb line must be sensitive, it must therefore be a so called residual line or "last line" (these spectral lines are those which with diminishing concentration are the last to disappear). The lines should, if possible, not display self-absorption because lines which do produce diminishing area density with the increasing element concentration. The relationship between area density and element concentration then becomes unclear. In shot range determination, however, only small amounts of Sb are present and self-absorption has little effect. The spectral lines should not be in the region of the cyanide (CN) bands (approx. 350 to

Fig. 33. Sections taken from spectrograms in the region of approx. 260 nm. The seven spectrograms show (from top to bottom) spectra with increasing shot range (notice that the reduction in the intensity of the Pb- and Sb lines and the uniformity of the Fe-line as an internal standard). The wavelengths are given in Å

460 nm). From all these considerations, the obvious choice falls on the Sb line 259.8 nm (Fig. 33).

The Fe line must satisfy the following conditions: the wavelength must be not be too different from the Sb line because the sensitivity of the photoplate is not the same all over, the gamma value is dependent on wavelength and moving the plate, even during photometry of plate, is inconvenient. It is also important that both lines have approximately the same excitation potential. Therefore the line Fe 259.9 nm is used as the reference line.

6.2.3 Sample Taking

According to the original method by Schöntag (1950) a piece of material (3 mm × 10 mm) is taken 1 cm from the middle of the entrance hole with either fine scissors or a punching tool. This sample is then placed in the cavity of the spectroscope electrode. It is advisable to take 4 samples from diametrically opposite locations and the mean of the four values taken. Care must be taken that the four locations are of equal distance from the center of the entrance hole.

A great advantage of this method of sampling and analysis is that the samples can be examined without any further preparation. The samples are therefore not contaminated by reprocessing (incineration) and much time and material are saved.

6.2.4 Carbon Electrodes, Excitation Conditions

The arc is generated with direct current of 4A. The open circuit voltage is 220 V. The sample-carrying electrode is the cathode, the counter-electrode is the anode. The cathode is cup-shaped and is of Quality 00 from the Ringsdorff-Werke (or the equivalent from other companies). The electrode carries the marking RW 0078 (Fig. 34). Using quality RW II, as put forward by Schöntag, has been shown to be unnecessary: this type has the added disadvantage of not always being available. They differ from one another in that they have different thermal conductivities (they consist of different carbon modifications). The counter-electrode, for reasons of economy, consists of a long carbon rod (3.6 mm diameter) RW 1. It need not be changed for each new analysis, it is only necessary that between analyses it is annealed.

The gap between the electrodes is 2.5 mm and this distance must be maintained. Due to the later quantitative evaluation and the therefore necessary photometry of the line, the spectral line must be homogeneous over its whole length. Because the spectral line is an image of the spectrograph slit, the slit must be illuminated homogeneously. The arc itself along its length from the cathode to anode is strongly heated and therefore radiates continuously, in the so-called cathode glow the metals are enriched (this is valid for all the metals which are of interest for our determination; not, however, for Hg, for example which glows over the whole length of the arc). If one allowed the light from the arc fall directly

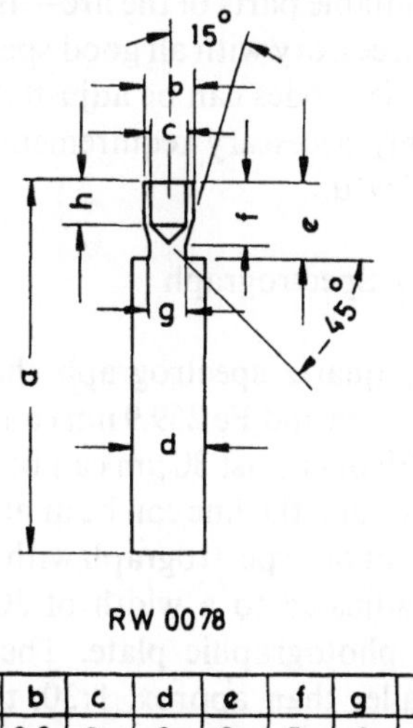

	a	b	c	d	e	f	g	h
mm	30	3,8	3	6	6	5	3	3,5
inch	1,18	.150	.118	.237	.237	.197	.118	.138

Fig. 34. A carbon electrode (cathode) from the Ringsdorff Co., Mehlem, FRG, Type RW 0078

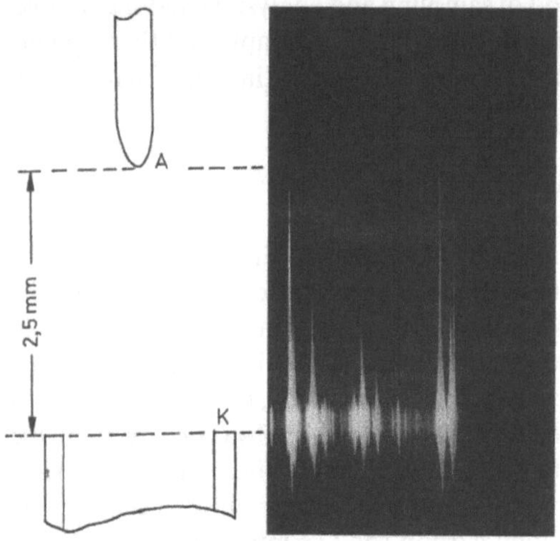

Fig. 35. Imaging of the arc on the slit. *Top:* anode (very hot = temperature radiator), *bottom:* cathode (maximum line intensity (cathode glow)

on the slit, one would obtain a spectrogram as shown in Fig. 35. Exact photometry of the lines is not possible because of their inhomogeneity. One must therefore homogenize the light falling on the slit.

At the same time, only that light from the arc should be allowed to enter the spectrogram, and therefore the photographic plate, which is emitted from the elements whose presence we are trying to detect and not, for example from the glowing anode. This light would fog the spectrum and reduce the detection sensitivity. This screening out of the useful light cannot take place at the arc because of the strong heating effect. One must therefore form an intermediate image (after Zeiss) and, using the image of the arc, mask out the useless light.

Both these requirements—uniform illumination of the slit and masking out of suitable parts of the arc—is fulfilled by a three-lens optical system supplied as an accessory with all good spectrographs. With help of the intermediate imaging, the electrodes can be adjusted vertically, horizontally or the gap set with ease: a highly necessary requirement when reproducible and exact results have to be achieved.

6.2.5 Spectrograph

Any quartz spectrograph that can adequately separate the spectral lines Sb 259.8 nm and Fe 259.9 nm can be used. The separation must be so great that a line width of at least 30 μm can be set up. This (geometrical) line width is necessary in order that the line can be analyzed spectrographically. To achieve this line width, the slit of a spectrograph with a collimator and camera of equal focal length must be adjusted to a width of 30 μm because the slit's image is produced 1:1 on the photographic plate. The aperture ratio should not be too small. When smaller than approx. 1:30, the sensitivity is reduced so that long, shot range determination becomes impossible. It is not just for sufficient separation of the Sb and Fe lines that the dispersion should be great but also because of reduced fogging of the photographic plate. With increasing dispersion, the background (the continuum of the spectrum) is reduced while the line density remains the same. The (weak) line that we are seeking becomes more easily visible against the background thus increasing the sensitivity and the range at which the shot distance can be determined is extended.

The evaluation of the lines follows the general principle set out in numerous publications and monographs e.g. Sellier (1967 and 1988).

6.2.6 Extending the Distance of the Shot-Range Determination

The spectrographic method has limits if the sample piece of material has such a small amount of soot elements that the detection by the emission spectrograph is no longer possible.

Two possibilities exist of extending the shot range which can be determined by overcoming the limitations due to the sensitivity of the method:

1. Taking a larger sample.
2. Using a more sensitive detection method.

The second possibility is realized either by AAS (Sect. 6.3) or NAA (Sect. 6.4). Using a sample having a greater area is not possible with Schöntags method because the sample will then not fit into the electrode. A limited increase in size is possible but the temperature of the electrode drops so much that quantitative vaporization of the soot elements no longer takes place. The volume of the sample must therefore be reduced in order that it can be placed in the cavity of the electrode. Concentration by the well known chemical methods has, however, the disadvantage that the sample can be contaminated, especially by lead, from the reagents used. The most favorable solution has proved to be incineration in a stream of air, because no reagents are necessary and therefore contamination by Pb and Sb is eliminated.

Sample taking—A circular ring with an external diameter of 50 mm and an internal diameter of 20 mm is cut out of the cloth under investigation. This ring is cut in half so that two samples are available.

Incineration—The sample is placed in a porcelain crucible which has been kept at a temperaturte of 850 °C over a long period and before this had lain in chromic acid. The crucible containing the sample is inserted into a quartz tube 1.2 cm in diameter that is concentrically enclosed in an electrical ring furnace. The annealing temperature to 550 °C and is controlled by a thermocouple converter. During incineration, an air current is passed over the sample. The incineration lasts about 20 min and no loss of Pb or Sb has been determined. The volume of the ash is so small that it is possible to put it in the cavity of the electrode used in the spectrographic method.

After the ash has been placed in the electrode 20 μg Sn in the form of a $SnCl_2$ solution is added. The Sn acts as an external spectrographic standard. For the "normal" spectrographic method, Fe is used as the standard because it is always present in textiles as a trace element, but it has considerable disadvantages. The vaporization temperature of Fe is much higher than those of Pb and Sb and so is not really suitable as an internal standard, whereas Sn behaves spectrographically like Pb and Sb.

This procedure can also be used for other methods (AAS etc.).

The *evaluation* of the lines and the calculations involved are carried out as previously described but this time with relation to the Sn line 259.4 nm. (This can of course be done in relationship to Fe but the results are not so exact).

Result: The area of the sample by the "normal" method in 0.3 cm^2, by this "improved" method approx. 33 cm^2 (half the circular ring). A concentration of soot elements of $33/0.3 = 110$ times higher is achieved. If we assume that the area density of the soot elements varies inversely to the square of the distance, then the increase in the range that can be determined increases by a factor of approx. ten. This means that for a particular cartridge-weapon combination, where the extreme determinable range is 60 cm, with the help of this enrichment the 60 cm is now extended by a factor of ten to 600 cm. The assumption that the reduction in soot element density follows the inverse square law is, however, wrong because the soot also loses intensity due to vortex formation and absorption. Therefore the reduction is not inversely proportional to the square of the distance but (for distances > 70 to 100 cm) to some higher power.

6.3 Atomic Absorption Spectrography (AAS)

6.3.1 General Comments

ESA as a procedure for determining the shot range is a proven method. It is simple and when a large sample for increased sensitivity is not required, no preparation of the sample is necessary and in addition it is a fast method.

Baumgärtner et al. (1963) introduced NAA for shot-range determination. This method is sensitive but requires time and effort. Results can only be obtained after a long period and the samples must be given to another agency.

The current method of choice is AAS. Kijewski (1974, 1981) and Sellier (1988) have had great success with this method.

6.3.2 Apparatus, Detection Limits

(For an introduction to the methodology of AAS, the standard work of Welz is eminently suitable.)

Because of the high sensitivity, the flame is not used for the metal detection but rather the graphite furnace technique. The increase in sensitivity reaches several orders of magnitude. The detection limits for the metals that concern us in the flame and the graphite furnace are compared in Table 9.

The most convenient, e.g. most sensitive spectral lines are to be found in the apparatus handbooks (e.g. Perkin-Elmer). Not only can the handbook be

Table 9. Detection limits for graphite furnace and flame (after Welz 1983)

Element	Wavelength [nm]	Detection limit in μg/l Graphite furnace*	Flame
Ba	553.6	0.04	2
Cu	324.7	0.02	1
Pb	283.3	0.05	10
Sb	217.6	0.1	30
Ti	365.3	0.5	50
Zn	213.9	0.001	1

* 100 μl solution

N.B. Zn should not be used as the chloride in graphite furnace technique as is stated in some "Cookbooks". The measurement cannot be reproduced and memory effects appear. As the nitrate, these mistakes will be avoided.

followed in this matter but also the temperature program for the graphite furnace can be used with practically no changes, at least for the first trials. Modifications are hardly necessary.

6.3.3 Taking the Samples, Preparation for Testing

We take a rectangle of 3 mm × 10 mm at a distance of .9 cm as is customary with the ESA method. The samples are placed in a 10 ml flask in 1 ml concentrated nitric acid and left on a hot-plate until they are completely dissolved (about 4 hours). The clear solution is diluted 1:10 with distilled water. From this dilute solution, 20 μl is placed in a graphite furnace (Perkin–Elmer) with a L'vov platform. The temperature programs for the most important soot elements are as follows: The inert gas is Ar (99.99%).

Zn = 213.9 nm Temperature curve in the furnace

	1	2	3	4	5	6
Temperature °C	120	400	1500	2600	20	
Ramp (s)	15	15	1	1	10	
Hold (s)	20	20	5	3	20	
				min. flow		

Pb = 283.3 nm Temperature curve in the furnace

	1	2	3	4	5	6
Temperature °C	80	120	500	2100	2150	20
Ramp (s)	2	20	15	0	1	10
Hold (s)	15	15	15	4	5	20
				min. flow		

Sb = 217.6 nm Temperature curve in the furnace

	1	2	3	4	5	6
Temperature °C	80	120	1400	2600	2650	20
Ramp (s)	2	20	15	0	1	10
Hold (s)	15	15	15	4	5	20
				stop flow		

Ba = 553.6 nm Temperature curve in the furnace

	1	2	3	4	5	6
Temperature °C	80	120	1400	2600	2650	20
Ramp (s)	2	20	15	0	1	10
Hold (s)	15	15	15	4	5	20
				min. flow		

A Zn-calibration curve for the 9 mm Luger SINTOX cartridge obtained by this method is shown in Fig. 36.

Kijewski et al. take two quartered rings (internal radius 0.9 cm and outer radius 4.7 cm from the middle of the entrance hole—an area of 5.3 cm^2) and incinerate them in apparatus from the Tracerlab Co. and with excited oxygen. The residues are dissolved in dilute HCl (in the case of Zn: HNO_3!) and examined in the graphite furnace.

The above mentioned authors encountered problems when measuring Sb (no other metals were investigated). They found that, in particular Ca- and Na-salts

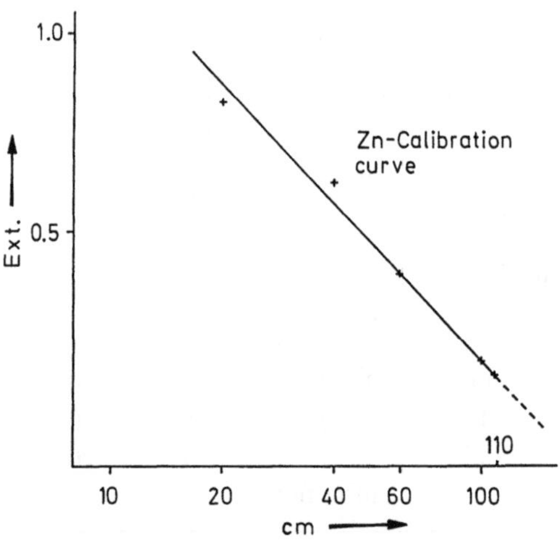

Fig. 36. Zn-Calibration curve for the 9 mm Luger cartridge with SINTOX primer. On the y-axis: Zn concentration in units of absorbance. For shot range determination a calibration in mass units is not necessary

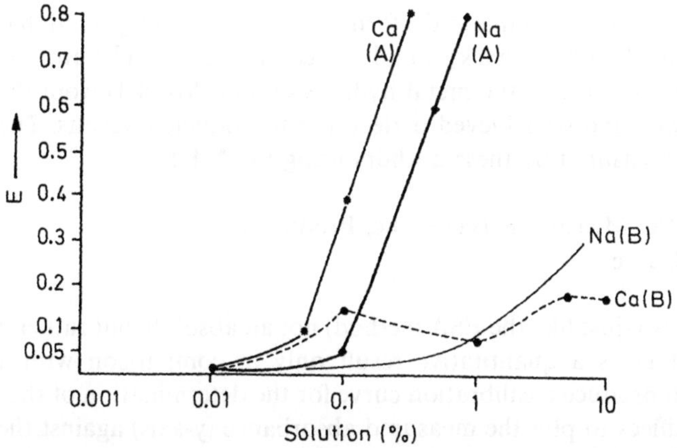

Fig. 37. Disruptions in the absorbance in the graphite furnace by highly pure salt solutions. Extinction (absorbance), A: measurement without background correction, B: measurement w background corrector, Na Na-chloride solution concentration in %, Ca Ca-salt solution (af Kijewski and Kleve, 1981)

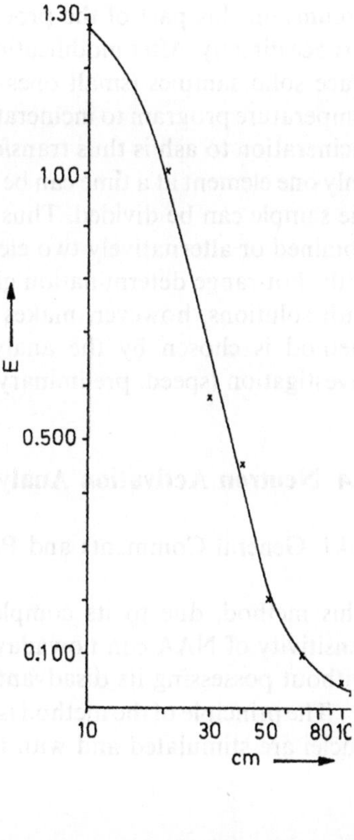

Fig. 38. Calibration curve for shot-range determination with antimony concentration as a function of the shot range. Weapon: Walter P38, ammunition: 9 mm Luger, brand: Geco (after Kijewski and Klewe 1981)

(blood soaked clothing) caused trouble in the form of a stronger background that could be partly remedied with the background corrector (Fig. 37). The authors therefore determined the Sb with the metal-hydride system (MHS 1) from the Perkin-Elmer Company and so achieved perfect and reproducible results. The curve in Fig. 38 was measured by these authors using the MHS.

6.3.4 Comments on the Measuring Technique, Producing a Calibration Curve

The AAS method is also (just like the ESA method) not an absolute but rather a relative method that gives a quantitative result only in comparison with a reference solution. To produce a calibration curve for the determination of shot range, however, it suffices to plot the measured absorbance (y-axis) against the shot range (x-axis), Fig. 37. It is only required that, during the whole measuring period, the values remain reproducible—a requirement easily fulfilled by modern equipment. To control the consistency of the results a calibration solution i.e. a solution of the element being investigated with an exactly known concentration is also tested. If the absorbance of this solution is constant at the beginning, in the middle and at the end of measurement series, one can be assured that the individual values in the series are correct.

Preparing the samples prior to their being measured in the furnace is inconvenient and time consuming. For this reason, there have been attempts to circumvent this part of the process and to introduce the solid sample into the furnace directly. After modification of the graphite furnace it has been possible to place solid samples (small ones) directly into the furnace and with a suitable temperature program to incinerate, to atomize, and to measure. The procedure of incineration to ash is thus transferred to the furnace. With this "direct" method only one element at a time can be detected. If the detection sensitivity is sufficient, the sample can be divided. Thus either two values for the same element can be obtained or alternatively two elements, e.g. Pb and Sb, can be measured which with shot-range determination gives increased certainty to the results. Working with solutions, however, makes it possible to test for many elements. Which method is chosen by the analyst is dependent on the requirements of the investigation (speed, preliminary tests, precision etc.).

6.4 Neutron Activation Analyses (NAA)

6.4.1 General Comments and Principle

This method, due to its complexity, has lost its significance. The very high sensitivity of NAA can nowadays be achieved with AAS or even improved on without possessing its disadvantages.

The principle of the method is that the sample is irradiated with neutrons. The nuclei are stimulated and with the emission of γ-quanta return to the original

Table 10. Nucleonic data for Sb

Nuclide	Neutron capture cross-section σ[barn]	Relative frequency %	Half-life τ	Emission [MeV]
^{122}Sb	6.8	57	2.8 d	$\beta-$ 1.99 $\beta+$ 0.57 γ 0.56(67%) γ 0.70 γ 1.26(4%)
^{124}Sb	2.5	43	60 d	$\beta-$ 1.48 γ 1.26
	0.03		≈ min	—

state (other possibilities: elastic scattering of the neutrons in the nucleus or neutron capture with the formation of a new isotope).

The relative data for Sb are presented in Table 10.

Sb belongs to the elements which can be detected extremely sensitively with NAA. Another member of this group is Ba. Ti and Zn, however, are less sensitive by two orders of magnitude and Pb cannot be detected at all by this method.

The most suitable nuclide for analytical determination is ^{122}Sb. It has an easily detectable γ-line of .56 MeV. In comparison to ^{124}Sb (half-life 60 d), it has the advantage of a shorter activation time and larger specific activity. The formation rate R of radioactive atoms with irradiation with neutrons is as follows:

$$R = N \cdot \sigma \cdot \phi \qquad (1)$$

N = number of atoms in the sample
σ = effective cross-section [cm^2]
ϕ = neutron flux in $n \cdot cm^{-2}/s$

The radioactive atoms decay at a specific rate which is dependent on the half life τ of the element. If a sample is irradiated for a long time t (in practice $t \geqslant 10 \cdot \tau$) and the formation of a new atoms equals the rate of decay, then

$$A^* = N \cdot \sigma \cdot \phi \,[\text{decays/s}] \qquad (2)$$

This is the highest activity, taking into consideration N, σ, and ϕ, that can be achieved. Because the number of atoms in the sample, N, and the effective cross section, σ, are given, the maximum activity, A* can only be increased by a higher neutron flux ϕ.

ϕ is specific to the reactor and varies between 10^{11} and $10^{13}\ n \cdot cm^{-2}/s$.

For reasons of speed, the production of the maximum possible activity, A* (and therefore the maximum sensitivity) is not sought after, since after a time $t = 2 \cdot \tau$, 75% of A* is achieved. For all that, Sb (half-life = 2.8 d) must be irradiated for 28 d (!) in order to achieve approximately A*.

The analytical relationship between the activity of the sample and the number of atoms contained therein is linear and therefore makes a quantitative analysis possible.

6.4.2 Method, Results

The general procedure of activation analyses is:

1. Cutting out of the sample from the target, in general, as circular rings or sectors.
2. Irradiation of the sample in a reactor with the irradiation time so that it is in agreement with the half-life of the element in question (see Eq. (2) in Sect. 6.4.1).
3. "Cooling off" of the irradiated sample for several days. This time is necessary for the radioactivity of other materials in the sample of diminish so that, in ideal circumstances, only the activity of the elements under investigation remains.

   The elements which can interfere with the method are, in particular Na and Cl. These elements come from the soaking in of blood in the proximity of the entrance hole or as general contamination (e.g. sweat) of the clothing. With a half-life of approx. 5 h, Na activity rapidly disappears. However, other elements present in biological material can also interfere; so that it can become expedient or even essential that the useful elements be separated from the interfering elements by radio-analytical methods and only then should the activity be measured. In the meantime, this has become a classical method, but for more details the reader is referred to Baumgärtner et al. (1963) and Krishnan (1967 I).

   Researching or testing a new method, the authors naturally, at the outset, work under optimum conditions. For example, as the target, filter paper is used which, from experience, we know is pure. Contamination is not present, so that a process of radio-chemical separation is unnecessary. The method seems to be rapid and elegant. A completely different picture becomes apparent when this "theoretical" method is put into practice in a "real" investigation where contamination is encountered. This is a factor that should always be kept in mind when judging a method of shot range determination. This remark is not just valid for NAA.

4. The radio-activity of the individual samples is then measured. With the aid of samples with a known content, the absolute quantity of the element can be determined. In the majority of cases it is sufficient to compare the activity of the actual shot relative to the activities of the calibration curve obtained from test shots.

Krishnan (1977 I) gave the following plan of the operation. He determined ^{122}Sb. The neutron flux ϕ was $10^{13}\,n\,cm^{-2}/s$, irradiated time 12 h. As target material, filter paper and cotton cloth were used. The measurements were made with circular rings, each one being .25 cm wide. The innermost ring had an

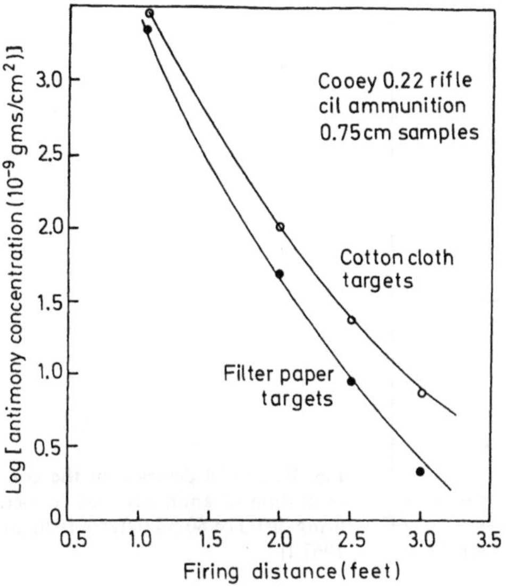

Fig. 39. Sb-Calibration curves for shot range determination (after Krishnan 1967 II)

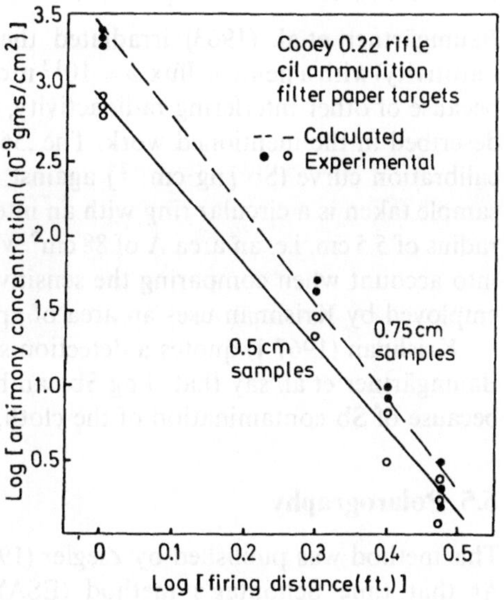

Fig. 40. As Fig. 39, target is filter paper. Influence of the different sample areas. Upper curve: inner ring radius $r_i = 0.75$ cm, lower curve $r_i = 0.5$ cm, width in both cases = 0.5 cm (after Krishnan, 1967 II)

internal radius of .5 cm. The measurement of the activity was made 4 or 5 days after activation (cooling-off period mainly due to Na). In Fig. 39 to Fig. 41 some results from this work are shown. The calibration curve derived from the results of individual rings obeys, according to Krishnan, the following equation:

$$\log(s) = 0.041 - 0.176 \cdot \log(\mathrm{Sb}) \quad (1)$$

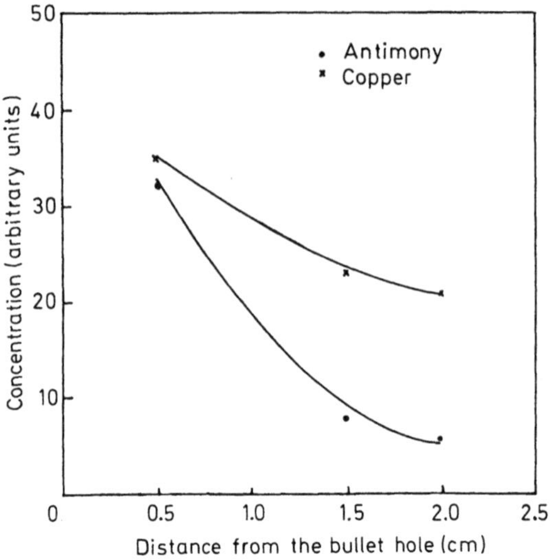

Fig. 41. Radial decrease of the concentration of antimony and copper, firing distance 60 cm (after Krishnan, 1967 I)

s = shot range in meters (from the original equation converted to the metric system;
Sb = Sb concentration in ng·cm^{-2}).

Baumgärtner et al. (1963) irradiated the cloth samples for 2 d (increase in sensitivity) with a neutron flux $\phi = 10^{13}$ n·cm^{-2}/s, waited one day and separated, because of other interfering radioactivity, the radioactive Sb. The separation is described in the mentioned work. The .56 MeV line of ^{122}Sb was measured. A calibration curve (Sb [ng·cm^{-2}] against shot range) is shown in Fig. 17. The sample taken is a circular ring with an internal radius of 1.5 cm and an external radius of 5.5 cm, i.e. an area A of 88 cm^2 which is very large. This must be taken into account when comparing the sensitivity with other methods. The method employed by Krishnan uses an area of approx. 1 cm^2.

Krishnan (1967 I) quotes a detection sensitivity of 1 ng Sb and 2 ng for Cu. Baumgärtner et al. say that .1 ng Sb can be sufficiently exactly determined but because of Sb-contamination of the cloth, this sensitivity cannot be used.

6.5 Polarography

This method was published by Ziegler (1965) and by Ziegler and Fehse (1966). At that time Schöntag's method (ESA) was in existence and shot range determination by NAA was being used for the first time. The authors claimed that polarography, with reference to its accuracy, was superior to ESA and that NAA was too costly in its instrumentation.

Polarography is convenient in that the two elements Pb and Sb can be tested for, side by side, in only one procedure.

Preliminary preparation: The pieces of cloth (1 cm in diameter) which have been punched out are placed in a quartz beaker containing H_2SO_4/HNO_3 and on the sand bath, at a temperature of 210 °C, turned to ash. The residue is placed

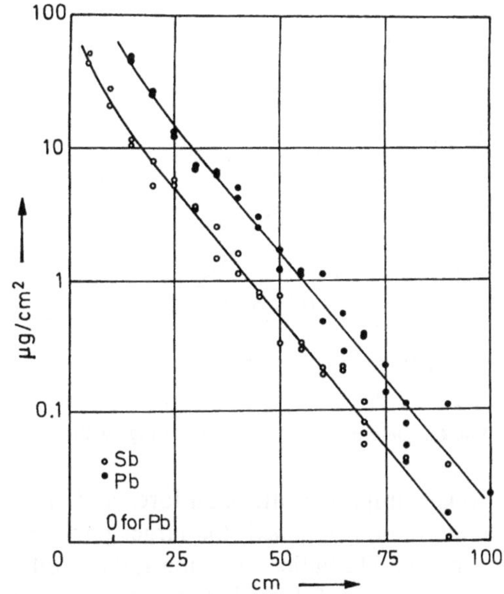

Fig. 42. Calibration curve for the polarographic method of shot range determination (from Ziegler und Fehse, 1966)

in HCl and warmed over a waterbath for 10 min, brought up to 5 ml and after deaerating with N_2 analyzed polarographically.

The scanning polarograph "Selector D" from the Atlas company with a maximum sensitivity of $2.5 \cdot 10^{-10}$ Å/mm. When cloth samples with Pb and Sb added were tested, detection limits of .025 μg/ml for Sb and .042 μg/ml for Pb were achieved.

The samples were taken at a distance of 1.5 cm from the center of the entrance hole. A calibration curve for a 9 mm Parabellum cartridge (Luger) shot from a Makarov pistol is shown in Fig. 42. The curve for Pb is displaced a little to the right for better visibility (notice the position "0 for Pb" on the x-axis). From this diagram it was possible to determine shot ranges of up to 90 cm (under favorable laboratory conditions)—admittedly from 70 cm upwards, triple the quantity of cloth was used.

When comparing the ESA and the polarographic methods one can see that both methods have a similar sensitivity. ESA, in its original form (without enrichment) is much simpler in the preparation of the samples. With reference to accuracy, there is little to choose between the two methods since most of the variations are not due to the methods but rather the material under investigation.

6.6 Other Methods of Shot Range Determination

6.6.1 Range Determination Using the Bullet-Wipe Ring

Schöntag (1957) proposed the use of the variation of the soiling of the bullet with the range to determine the distance. The principle behind this is simple. The bullet accumulates soot and soot elements on its passage through the barrel. On leaving the barrel, the bullet jacket is soiled the most. The bullets rotation (approx.

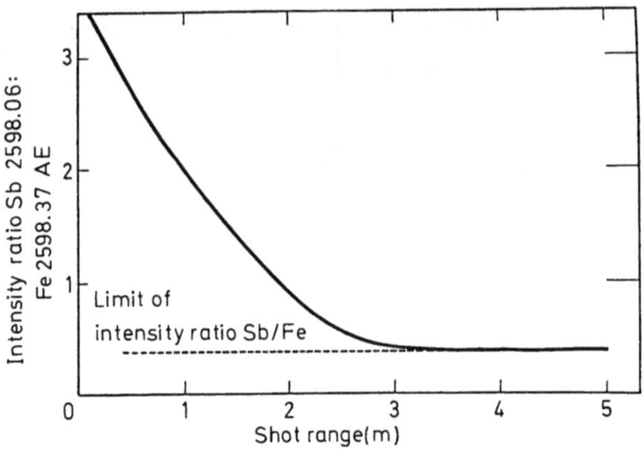

Fig. 43. Determination of the shot range with use of the bullet wipe ring after Schöntag (1957)

3000/s) imparts centrifugal force and cleans the jacket during the flight until a limit value is reached (the larger particles fly off first, the smaller ones remain clinging). The bullet-wipe ring at the point of entry is approximately proportional to the soiling of the bullet and can therefore (according to Schöntag) be used for shot-range determination.

Schöntag investigated the Sb content of the ring using ESA. As reference element, he too used the Fe content of the material. He used a pair of spectral lines Sb 259.81/Fe 259.84 and formed the quotients from the intensities i.e. the difference in blackening of the two lines. A calibration curve from his work is shown in Fig. 43. The outer limit of the range that can be determined is about 3 m.

I have, myself, experimented with various weapons and ammunition. The values diverge so greatly that with this (theoretically elegant) method it does not seem possible to determine the shot-range reliably. A possible and probable source of error is that the spectrographic method is a relative method, the Sb (Pb, Ba) is determined relative to the area density of the Fe in the material. With the ESA method, however, a well defined area is taken (with a constant amount of Fe), whereas with this method a piece of cloth is taken partly destroyed by the bullet and whose size therefore varies. It is clear that the absolute quantity of Fe

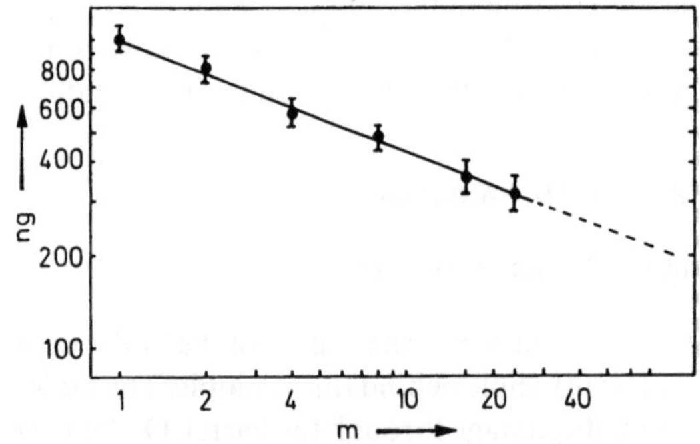

Fig. 44. Determination of shot range from the bullet-wipe ring using AAS (Kijewski and Lange, 1974). Sb content in the bullet wipe as a function of range

also varies and this variation in the reference element explains the scattering of the measurements.

More recently, Kijewski and Lange (1974) worked on the same problem by using AAS instead of ESA.

Method: For each firing distance (1, 2, 4, 16 and 24 m), 10 shots were fired (Walther P38 pistol, 9 mm Parabellum (Luger)), the bullet-wipe ring was removed with a punch with 16 mm diameter, and ashed. The residue was absorbed with 400 μl 2n-HCl from which 20 μl was injected into the graphite furnace. The temperature program: 40 s at 100 °C—dried, 60 s at 750 °C—thermally decomposed, and 30 s at 2500 °C—atomized, the buffer gas was Ar. The Sb line 231.2 nm was measured. The result is shown in Fig. 44. The realization of the calibration in absolute quantities is not given.

From the statistical analysis of their measurements, the authors came to the following conclusions: a shot from a range of 16 cm can be distinguished with 99% certainty from a shot from a range of 2 m. Between 4 and 8 m the scattering of the measurements from one distance overlap that from the other, therefore giving no possibility of differentiating the two. It could be shown that up to range of 24 m the Sb-content of the bullet wipe ring had no constant final value (Fig. 44). Possible errors could be due to varying strength of the textiles e.g. at the seams. It is a matter of experience that in areas where the cloth is stronger that the wiping of the bullet is more complete and the ring is more intensive.

Every expert knows that the intensity of the bullet wipe ring can vary from shot to shot and that the intensity is dependent on the number of shots. This is not apparent when inspecting the bullet wipe ring. It is unique and therefore incapable of being repeated.

Generally speaking, the method cannot be recommended. Only in some cases may it be useful for making a rough estimate of the shot range, e.g. it can succeed in proving that the shot range was more than 100 m (Kijewski).

6.7 Possible Errors in Shot Range Determination

In general, the presence of soot elements around the entrance hole is interpreted as powder soot blackening and according to intensity assumed to show an intermediate range shot from a definite distance. Unfortunately, apart from the usual possibilities of contamination (e.g. from the person's job), there are other occurrences which can cause the soot elements Pb and Sb, from sources other than the soot (e.g. lead from the bullet) or from the soot by different paths, to be deposited around the entrance hole. Some of these phenomena will now be described.

6.7.1 The Vinogradov Phenomenon

Beyond a certain shot range, no more powder particles and no more powder residues can be found on the target. This distance which is the upper limit of an intermediate range shot, is dependent on the weapon, the ammunition and the method of investigation. With the latest method the zone of intermediate range shot has been extended to approx. 2 to 3 m. Vinogradov and also Movshowich, however, made the following observations. From the detection of Pb around the entrance hole by using contact-chromatography (Leszczinski) they concluded in an actual case that an intermediate range shot was being investigated. This conclusion contradicted all the other evidence which indicated that the shot range had been at last 150 cm. The authors after numerous experiments made the following statement:

1. When shooting into a two-layered target, deposition of metal can take place on the second layer.
2. Depending on the distance between the two layers, the form of deposition changes: close together gives dust-like deposits, further apart the deposits are spotted.
3. The changes correspond to those from an intermediate range shot despite being from a much greater distance.

This phenomenon is due to the following (see Fig. 45): as the shot passes through the first target, the bullet produces a bullet-wipe ring on the edge of the hole (it consists of oil, powder soot, and soot elements). If this target is brittle enough, pieces from the edge are broken away and strike the second target, taking with them parts of the bullet-wipe ring. Therefore, the second target also shows signs of powder-soot blackening (presence of soot elements Pb, Sb, Ba) although these elements do not come directly from the powder soot, i.e. not directly from the muzzle to the target (genuine powder-soot blackening), but rather via the bullet respectively its bullet-wipe ring. This is therefore a pseudo-powder-soot blackening, which must not be confused with powder-soot blackening from an intermediate range shot. It is important to note that in the Vinogradov phenomenon all the soot elements in the pseudo-powder-soot blackening come from the primer, just as the normal blackening does. This is in contrast to the following phenomena where the elements Pb and Sb originate from the material from the bullet.

It is not just from the bullet wipe ring that a deposition of soot elements on the secondary target can take place. Abrasion from the bullet's material can, under certain circumstances, also cause elements to be found on the second target. This is especially so in the case of pure lead bullets because of:

1. Due to the softness of the lead a relative large amount is abraded.
2. This abraded material contains the same elements (except Ba) as the primer, i.e. Pb and Sb (Sb is alloyed with Pb to harden it).

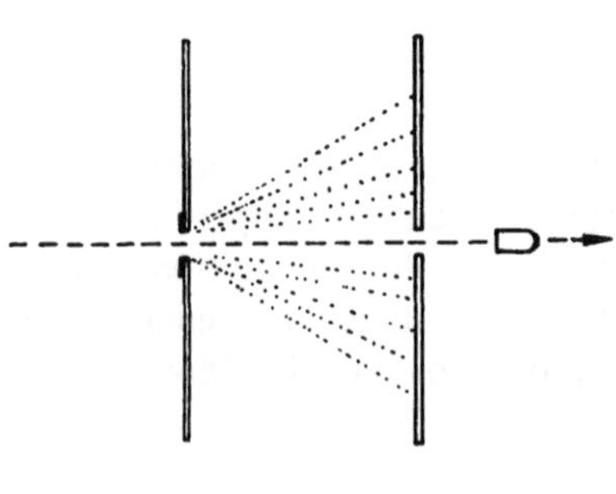

Fig. 45. Schematic representation of the Vinogradov phenomenon. See text for further details

An uncritical interpretation of the presence of Pb and Sb can lead to the diagnosis of an "intermediate-range shot" when it is, in reality, a long range shot. The quantity of Ba, (if Ba is present in the primer) can clarify the situation, but so can an inspection of the actual scene to reconstruct the flight path of the bullet in order to see if there were any intervening targets.

An abnormally small penetration depth of the bullet would also speak for a perforation of an intermediate target or a considerable ricochet ("face" of the bullet: bullet deformation?).

It is a prerequisite for the existence of this phenomenon that there are two targets not too far apart with the primary target being hard and brittle so that the described process can take place. The soot elements can be found on the secondary target (possibly a person). If the primary target is soft, e.g. cardboard, then the Vinogradov phenomenon cannot be produced as I have shown with my own experiments. Kijewski (1974) experimented with glass plates as the primary target. Especially from shots using pure lead bullets (.22) intensive deposition was found on the secondary target.

This effect is understandable: pure lead bullet are not just wiped of their jacket soiling (powder soot causing bullet—wipe rings), but the hardness of the glass sheet brings about abrasion of the bullet material that is then crushed into the secondary target giving the appearance of an intermediate range shot.

If there is a suspicion of this effect an investigation should be undertaken to see if there is any Ba—if this is contained in the primer. Ba must then, in the case of a close-range shot, be contained in addition to Pb and Sb (it must be determined to what extent the existence of Ba can be proved in connection with the range of the shot).

With SINTOX ammunition mistakes are not possible, since soot metals (Zn and Ti) and bullet elements (with lead bullets, Pb, Sb) are not identical. On the other hand, jacketed bullets have alloys containing Zn in the bullet's brass jackets.

In short the Vinogradov phenomenon can lead to the false diagnosis: "intermediate range shot" as being one of long range (distant shot).

6.7.2 Back-Scattering Effect

There is a certain similarity between the just mentioned phenomenon and this effect first described by Messler and Armstrong (1978). When two sheets are placed in line and the distance between is not too great (Fig. 46a), the first sheet being soft, e.g. cardboard, and the second hard, e.g. glass (the reverse of the Vinogradov phenomenon) one can see on the reverse side of the cardboard a distribution of blackening that with a cursory glance could be taken to be powder soot blackening. (Fig. 46 b + c).

Explanation: the pure lead or partly jacketed bullet flies through the soft target, then through the hard one. There, from the forcing in and through, a great amount of heat is produced. Bullet material vaporizes or is abraded and is scattered backwards from the direction of the shot onto the reverse side of the primary target. The black-gray deposit consists of bullet material (Pb and Sb).

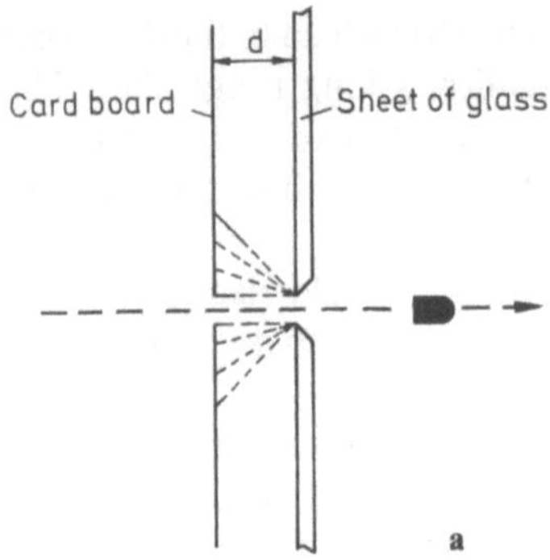

a

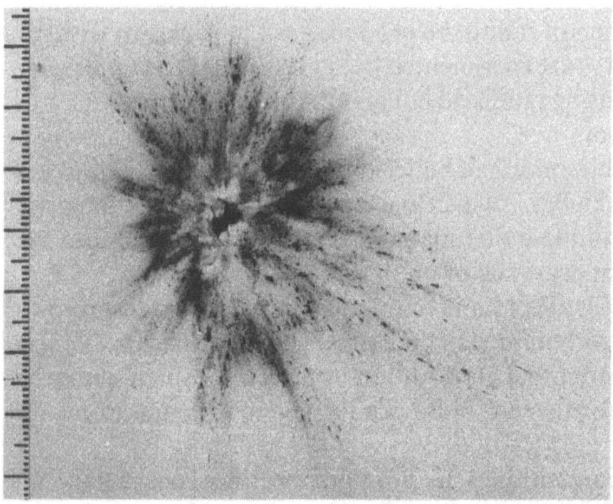

b

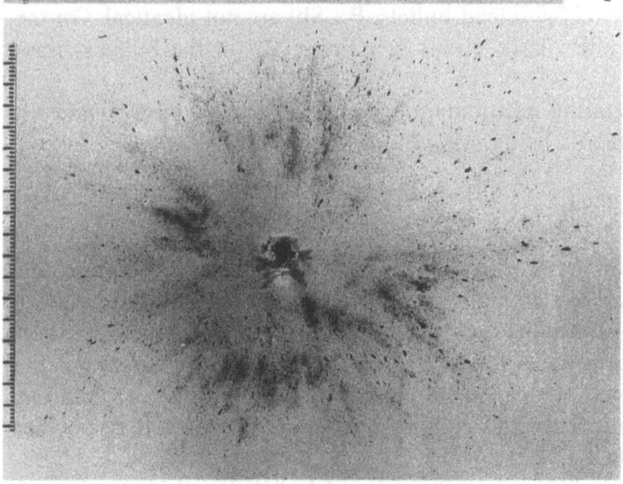

c

Fig. 46 a–c. Pseudo-powder-soot blackening (after Messler and Armstrong)
a) Experimental device: 1. Target: cardboard; 2. Target: sheet of glass 4 mm thick with a separation d; small calibre pure lead .22 l.r., $V_0 \approx 330$ m/s. **b)** d = 2 cm. **c)** d = 4 cm. In each case the side of the cardboard which faced the glass sheet is shown e.g. the side away from the muzzle

Under strong magnification, the fused droplets can be seen. The effect is not dependent on the range. It is greater when the bullet's energy is higher and the secondary target is harder. A decisive factor is also the distance between the two targets. With increasing separation the blackened area increases rapidly with the density decreasing by the same proportion.

In Fig. 46b + c two such "powder-soot-blackenings" are shown, with the exception of the separation *d* all the parameters where kept constant.

Confusion with an intermediate range shot (identification of "powder soot blackening") from the reverse direction (blackening on the reverse side of the primary target) should not cause any trouble because an appropriate (hard, brittle) secondary target must be in close proximity to the primary target. If the secondary target is a sheet of glass it is possible that (when the scene is examined) the glass has fallen out the frame and is no longer in the vicinity of the primary target.

Close scrutiny, however, reveals that there are differences between real and pseudo-powder soot blackening so that confusing one with the other is hardly possible in spite of not knowing of the presence of a secondary target:

1. The pseudo form possesses a distinct "centrifugal" structure. It appears as if all the particles have come out of the same bullet hole.
2. The larger particles (see Fig. 46 b + c) do not consist of powder particles but of bullet fragments.

6.7.3 "Powder Soot Blackening" from Fragmentation of Bullets ("Ghost Powder Soot Blackening")

High velocity bullets can break up even when penetrating soft tissue and even when they are full-jacketed bullets, e.g. .223 Rem. It is then obvious that the especially hunting bullets, whose front section disintegrates while the rear section remains complete and produces, or should produce, an exit hole. Irregular, stronger disintegration takes place when these bullets strike bone.

In all these cases, especially from the last mentioned—apart from large and small splinters—very fine lead particles can be formed which together with splinters fly out of the exit wound. In this way blackening of the clothing (at the exit hole!) is produced from the inside. The positive identification of "powder-soot blackening" can lead to the false diagnosis, "entrance hole and intermediate range shot", even though it is in reality an exit hole ("Geister-Beschmauchung" = "Ghost powder soot blackening", after Hoffmann, BKA).

The Pb and Sb originates (to recapitulate) from the core of the bullet with Sb being part of the lead alloy. Generally speaking, the relationship between the quantities of Pb and Sb is different in the bullet to that in the primer. Therefore by measuring Pb/Sb we can determine the origin of the lead. If, in addition, Ba is found, it can be definitely said that this is true powder-soot blackening and the Pb comes from the primer because barium is an ingredient of the primer and not of the bullet.

A "pseudo-bullet-wipe ring" can also be present at the exit hole. The edge of the hole has large quantities of Pb and Sb present (from the lead of the bullet!) which, because of the mechanism of the deposition, is understandable. With dark cloth one cannot check by sight if a (typical) bullet-wipe ring is present or not. In general one does not fall blindly into the trap and make a false diagnosis, i.e. confuse entrance and exit holes. Finally, with shot that goes right through, both holes would be investigated (or more exactly, the edge of the holes) for the presence of a bullet wipe ring. When, in exceptional cases, two such rings are found, the examiner becomes immediately suspicious. He searches for signs which show differences between them (this, as such, sure sign the "bullet wipe ring" here can leave the expert in the lurch). The shape of the opening must be examined. In most cases it is easy to discriminate the entrance from the exit hole. The entrance hole is the size of the caliber and exit much larger (due to the exit of the bullet and bone fragments as well). However, there are cases in which all the splinters of the bullet remain in the body and only a large piece of the bullet emerges (this type of bullet is preferred by hunters) producing a caliber sized exit hole. The morphological findings do not allow safe conclusions to be made from a entrance hole. In this situation, it must be remembered, that bone fragments produced by the bullet will travel in the direction of the exit wound, and bullet splinters also have this tendency. It is therefore good policy, when presented with a shot from a HV-bullet (full- or semi-jacketed) to take an X-ray picture before the autopsy.

It should be emphasized that this "Ghost powder soot blackening" can only come from HV-bullets with high energy (> 1000 J).

6.7.4 Interference with the Detection of Soot Elements Due to Various Physical and Chemical Effects

Kijewski (1977), among others, has done comprehensive work on this problem.

1. Detectability of the soot element Sb after the action of water on the target (clothing). The results are shown in Fig. 47. The experimental conditions are set out in the figure legend (the piece of material was shot with a .45-revolver at a range of 40 cm).
2. Detectability of Sb on shot-affected skin after being exposed to fire (Fig. 48). Even after exposure to fire (from a bunsen flame, the temperature was measured 2 mm under the skin with a thermocouple), Sb shows a remarkable persistence. This may partly be due to the fact that Sb, after the shot, is not only present on the surface of the skin but also in deeper layers, as shown in Fig. 49. The Sb-distribution reaches a depth of approx. $\frac{1}{4}$ mm.

   With experiments on burning with large quantities of leaded gasoline, large quantities of Pb and traces of Sb wer even found on skin not affected by the shot. This is undoubtedly from the lead content of the gasoline.
3. Even with washing of the clothing, great persistence of Sb was evident (Fig. 50). In spite of being treated with HNO_3, 10% of the original amount of Sb was still

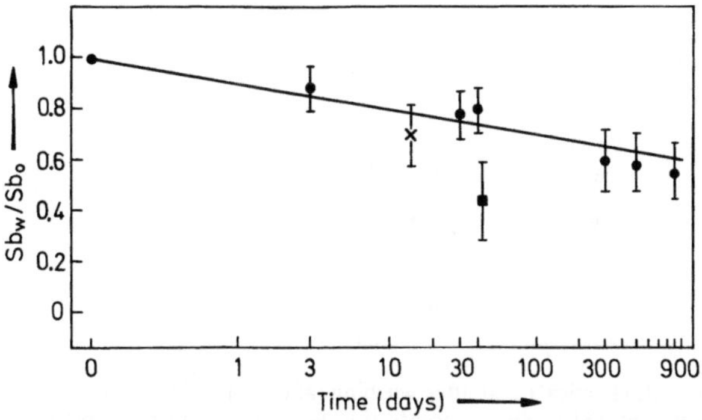

Fig. 47. The reduction of Sb-concentration with irrigation time (from Kijewski 1977). $[Sb]_w$—Sb concentration after irrigation; $[Sb]_0$—Sb concentration without treatment in pieces with the same area. ●—soaking in still tap-water; x—soaking after 14 d in river water; ■—after 42 d in flowing tap-water

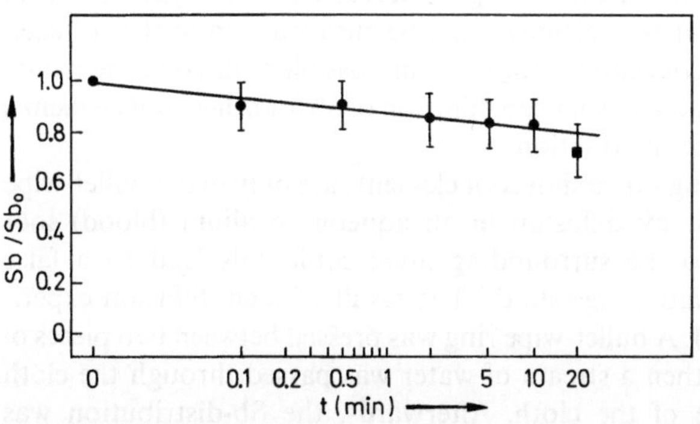

Fig. 48. Detectability of Sb in the skin after treatment with flame (from Kijewski 1977) Temperature measure at a site 2 mm below the surface of the skin: ●—300°C; ■—500°C

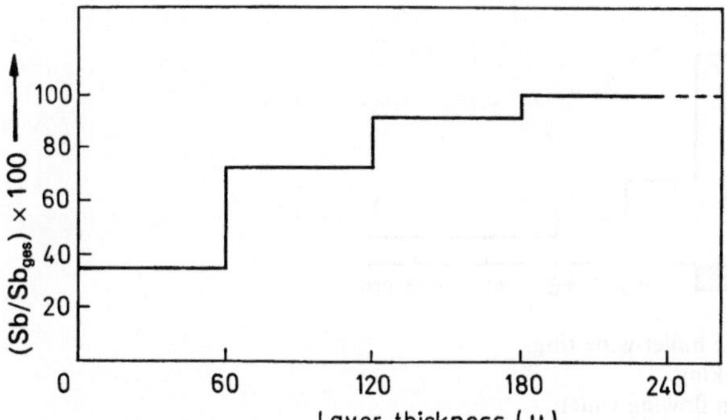

Fig. 49. Sb-distribution at various depth in the skin (from Kijewski 1977). Shot range 20 cm. Weapon 9 mm pistol

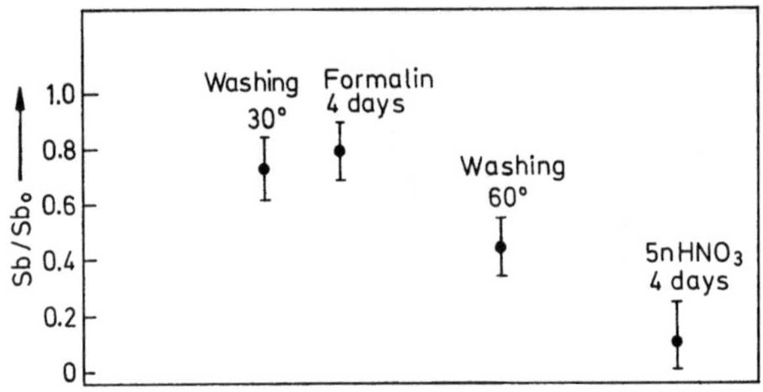

Fig. 50. Sb-losses after treatment with chemicals and after washing (after Kijewski 1977) [Sb] Sb—concentration after treatment; $[Sb]_0$ Sb—concentration without treatment in pieces of equal area

there. Kijewski was of the opinion, that Sb was present in the weave in an sparingly soluble form. Intermediate range shots can cause the synthetic fibers to melt and bake so that soot elements can be encapsulated in the softened fibers so that even with concentrated acid it is impossible to dissolve them out.

4. It is an interesting question how far can diffusion of Sb (or other soot elements) change the soot element distribution.
   We know that with a long range shot soot elements are only in the bullet-wipe ring. Is it possible that by diffusion in an aqueous medium (blood) soot elements may travel to the surrounding area; could this lead to a false diagnosis of "intermediate range shot"? The result of such diffusion experiment is shown in Fig. 51. A bullet-wipe ring was pressed between two pieces of cloth and soaked and then a stream of water was passed through the cloth parallel to the surface of the cloth. Afterwards, the Sb-distribution was measured. After being soaked, Sb was only found in the contact zone between

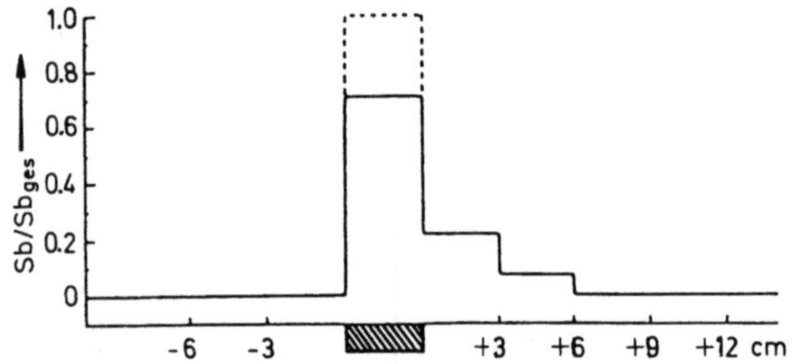

Fig. 51. Diffusion of Sb from the bullet-wipe ring.
——— Sb-distribution after soaking.
– – Sb-distribution after being in flowing water.
/// Contact zone between ring and cloth.
–x Distance from the contact zone (cm) against the current.
+x Distance from the contact zone (cm) with the current

the ring and the cloth, to some degree a "printing". In contrast, after the stream of water had been passed through, no Sb was found in the direction from which the water came; in the contact zone 70% of the total amount found by "printing"; the missing amount (30%) had traveled in the direction of the current, the quantity diminishing with the distance from the source. This could have a significant effect on wounds which bleed freely.

5. Finally, Kijewski investigated the influence of putrefaction on the detection of Sb. In samples of skin with shot wounds which were putrefying and partly decomposed 3/4 of the original amounts of Sb was still there.

I have, myself carried out experiments with the aim of determining if, after 3 days lying in still water, the soot elements are washed away or displaced. The result was that from the bullet wipe ring no soot elements contaminated the surrounding area. The powder soot blackening in textiles which had been lying in water was marginally less than textiles which had been kept dry when they had been shot from the same range.

The experimental results quoted here are, of course, dependent on the type of weapon and ammunition and on the characteristics of the target. The results cannot therefore be transferred unconditionally, but they do demonstrate an astonishing resistance to physical and chemical influences. This means that one should attempt the detection of soot elements even in apparently hopeless cases. At least one could give an estimate of the shot range.

6.7.5 Changes in the Pattern of Powder Soot Blackening Due to a Muzzle Silencer

Silencers are used to reduce the noise of the muzzle blast (for an overview see e.g. Sellier 1982). They are screwed onto the muzzle. Their exhaust-gas openings are

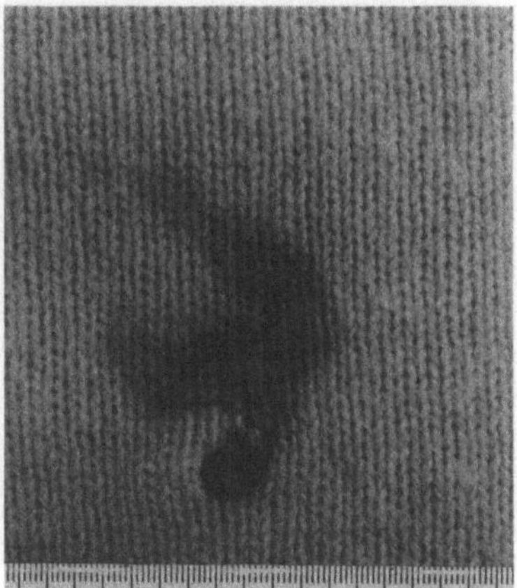

Fig. 52. Contact shot from a small caliber rifle (self-loader with silencer whose exhaust gas openings are on the upper-side of the casing)

either placed symmetrically to the axis of the barrel (in the simplest case a concentric hole through which the bullet also emerges) or they are supplementary openings in the casing of the silencer. In every case, the powder soot blackening of the target is reduced and/or modified.

With symmetrical silencers the radial symmetry of the blackening is maintained, and with the asymmetrical ones it is asymmetric (Fig. 52). The latter case is more convenient for the investigator because the shape of the blackening makes it obvious that a silencer has been used. In the first case, however, the intensity of the blackening is reduced but the shape remains the same (Fig. 53). If the silencer is not taken into consideration this can lead to determination of a longer shot range.

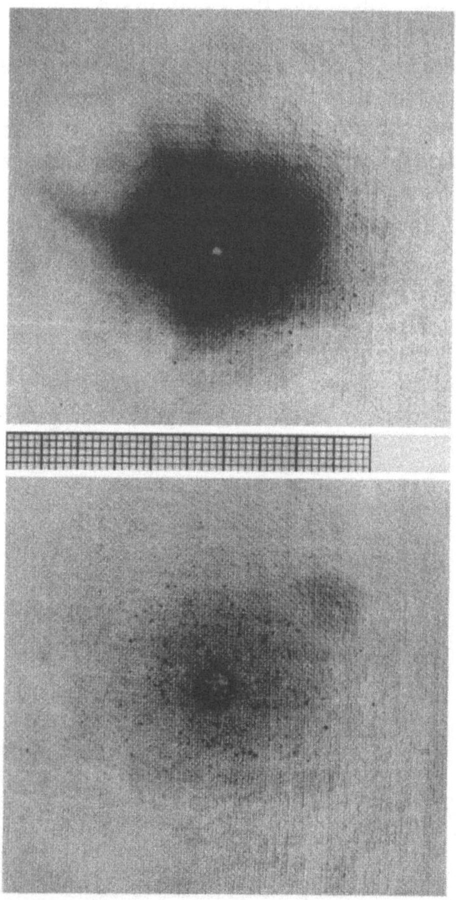

a

b

Fig. 53 a, b. Effect of a muzzle-silencer on the powder soot pattern. Small caliber sporting gun, shot range 20 cm, **a)** without; **b)** with silencer

7 Shot Range Determination for Shotguns

7.1 General Remarks

The range that can be determined for shotguns is much greater than that of shots from "normal" cartridges which only contain a single bullet. With a single bullet the intermediate range shot indications (powder soot and powder particles), because of the low cross-sectional capacity are restricted to 1 meter range, whereas with a shot gun, the density of pellets in the target is used to determine the range. It is thus possible in favorable cases to ascertain a shot range of more than 30 m. The decisive factors are the shape and type of pattern and its changes as a function of the shot range, the type of barrel, and the characteristics of the shotgun shell. These dependences will now be discussed.

Numerous tables and diagrams on the connection between the size of the hit pattern and the range have been published in hunting literature (e.g. Lampel). Since for hunting purposes, only ranges of 20 to 50 m are interesting, these tables only concern themselves with these distances. Extensions of this to closer ranges or to cartridges other than those hunting (e.g. .22, .38 or .44 caliber i.e. cartridges for handguns) have been published by many authors. Wray et al. (1983), Rowe and Handson (1985), Chugh (1986), Fann et al. (1966), Speak et al. to name but a few. More theoretical discussions of radial distribution of shot pellets originate from Nag and Lahiri (1986).

When studying this literature, however, one should keep in mind that the quoted accuracy is hardly ever reached, if at all, due to the "capriciousness" of shotgun shots.

7.2 Characteristics of the Barrel, Choke

The type of bore in the vicinity of the muzzle determines how the shot charge is scattered and the resulting density of the impacts on the target (density = number of impacts per unit area). This part of the barrel is rather constricted in comparison to the rest (Fig. 54). This construction is called the choke and holds the shot charge together. There are various types of bore:

1. Pure cylinder: the caliber is constant the hole length of the barrel,
2. Improved cylinder: the caliber is reduced uniformally along the whole length of the muzzle,
3. Weak choke: ($\frac{1}{4}$ to $\frac{1}{2}$ choke),
4. Medium choke ($\frac{1}{2}$ to $\frac{3}{4}$ choke),
5. Full choke.

The measure of how much smaller the narrowing of the barrel is than the cylindrical part of the barrel according to Lampel Mahrhold is:

full choke	between 1.0 and 0.75 mm; normally 0.875 mm
$\frac{3}{4}$ choke (improved modified choke)	between 0.35 and 0.875 mm; normally 0.76 mm
$\frac{1}{2}$ choke (modified choke)	between 0.35 and 0.5 mm; normally 0.45 mm
$\frac{1}{4}$ choke (improved cylinder)	mostly 0.25 mm.

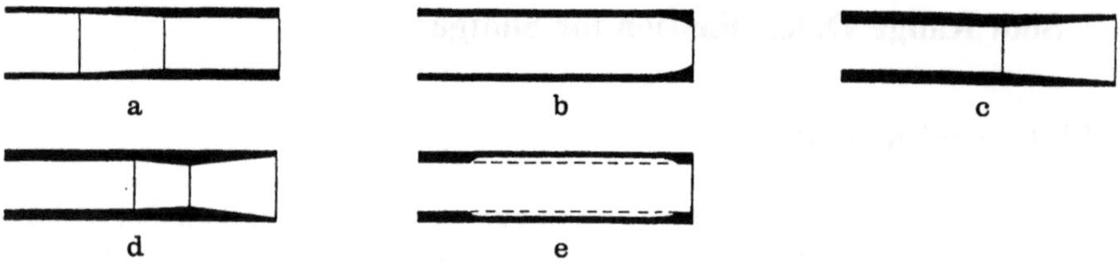

Fig. 54 a–e. Various forms of chokes in shot guns.
a) Normal choke. **b**) Ogival choke. **c**) "Bell" choke. **d**) Skeet choke No. 1—is used to produce a rapid opening-out of the charge in a short distance **e**) Recess bore, for incorporating in cylindrical barrels.

How strongly the scattering of the pellets is affected by the choke is shown by high speed photography. Figure 55a and b are photographs of the shot column from a shot out of a cylindrical barrel, i.e. without choke, and from a barrel with the fullest choke at a distance of 1.20 m from the muzzle 56a, b, and c shows the same at a distance of 3.6 m from the muzzle (Photographs from the Laboratory of the Winchester Rep. Arms Co. by Lampel). The cylindrical object is the felt wad which is situated between the propellant and the charge of the shot in order to prevent the propellant gases from swirling through the pellets during the passage through the barrel. This turbulence would scatter the pellets. Although the felt wad fulfills this task it has the disadvantage that it flies into the rear part of the charge of shot and tends to scatter it (Fig. 55a). The affected pellets are knocked from their normal trajectory and strike a long way from the middle of the shot pattern and from the so-called fliers (Fig. 57c and e). Another effect produces strong scattering: outer pellets from the charge are ground by the hard inner surface of the barrel and thus become unsymmetrical. They are also forced out of the true trajectory, this time by wind-resistance, and often end up a long way from the normal shot pattern. Both these effects can be avoided by new development from the USA (see Sect. 2.1.2).

The shot charge is held in a plastic cup which prevents contact between the pellets and the barrel. This cup when it leaves the muzzel becomes so distorted that it is freed from the charge and no longer has any effect on it. In Fig. 57a–e the relationships are shown. The plastic cup holding the charge of shot and the laminated wad are in one piece ("power piston"). The lamination gives additional cushioning. A felt wad allows almost the full blast of the gas pressure to get through to the charge of shot causing some of the pellets to clump together whereas the cushioning effect of the laminated wad produces a "gentle" acceleration.

The influence of pellet deformation on the spread of the charge of shot has been shown by Thornton, Crim and Rios (1987). They shot normal shotgun cartridges 12/70 with a pellet size of approx. 3 mm in diameter in the first series of tests and thereafter measured the hit pattern. In a further series the pellets were first removed from the cartridge and flattened on both sides and then replaced in the cartridge. For the normal cartridge at a specific range a hit pattern with a maximum diameter of 48 cm was measured. The altered cartridge, however, gave

Fig. 55 a, b. Influence of the choke on the radial and longitudinal spreading of the shot charge (after Lampel-Seitz). **a**) Shot from cylindrical bore, 1.2 m from the muzzle. **b**) Shot from a barrel with the strongest choke, 1.2 m from the muzzle

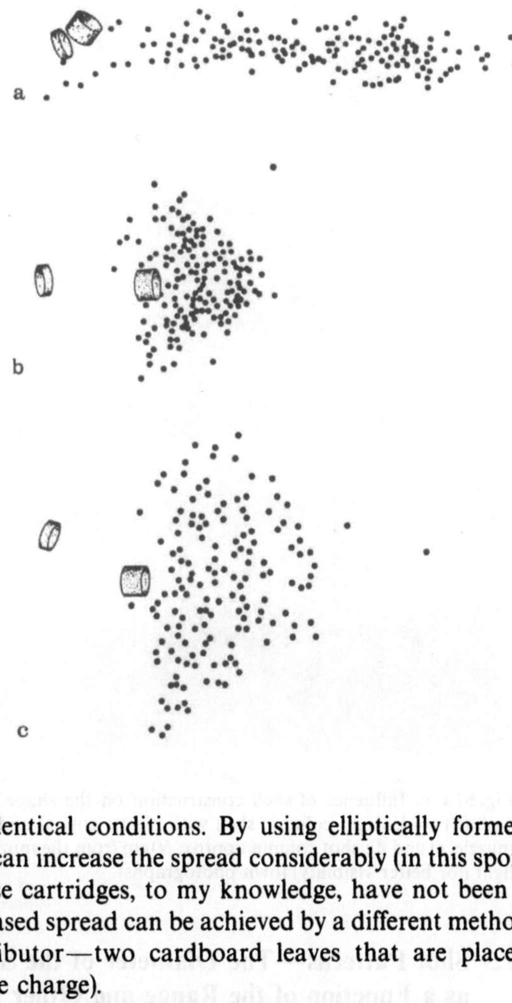

Fig. 56 a–c. Influence of the choke on the radial and longitudinal spreading of the shot charge (after Lampel-Seitz). **a**) As in Fig. 2b, but 3.6 m from the muzzle. **b**) Shot from barrel with a modified choke, 3.6 m from the muzzle. **c**) As in Fig. 2a, 3.6 m from the muzzle

a diameter of 70 cm under identical conditions. By using elliptically formed pellets in skeet cartridges one can increase the spread considerably (in this sport this is a desirable effect). These cartridges, to my knowledge, have not been a commercial success. The increased spread can be achieved by a different method (by the so-called pellet distributor—two cardboard leaves that are placed together to form a cross in the charge).

The degree of unfolding of the plastic cup can also be used to determine the range (Sect. 7.4).

In brief, the density of pellets (at the same range) depends decisively on the choke of the barrel. "Fliers", due to new kinds of shot shell construction are, to all intents and purposes, prevented.

The deformation of the pellets, whether intended (through special treatment of the pellets) or undesired (due to the abrasion in the barrel or due to rapid acceleration) has a considerable influence on the pellet density in the target.

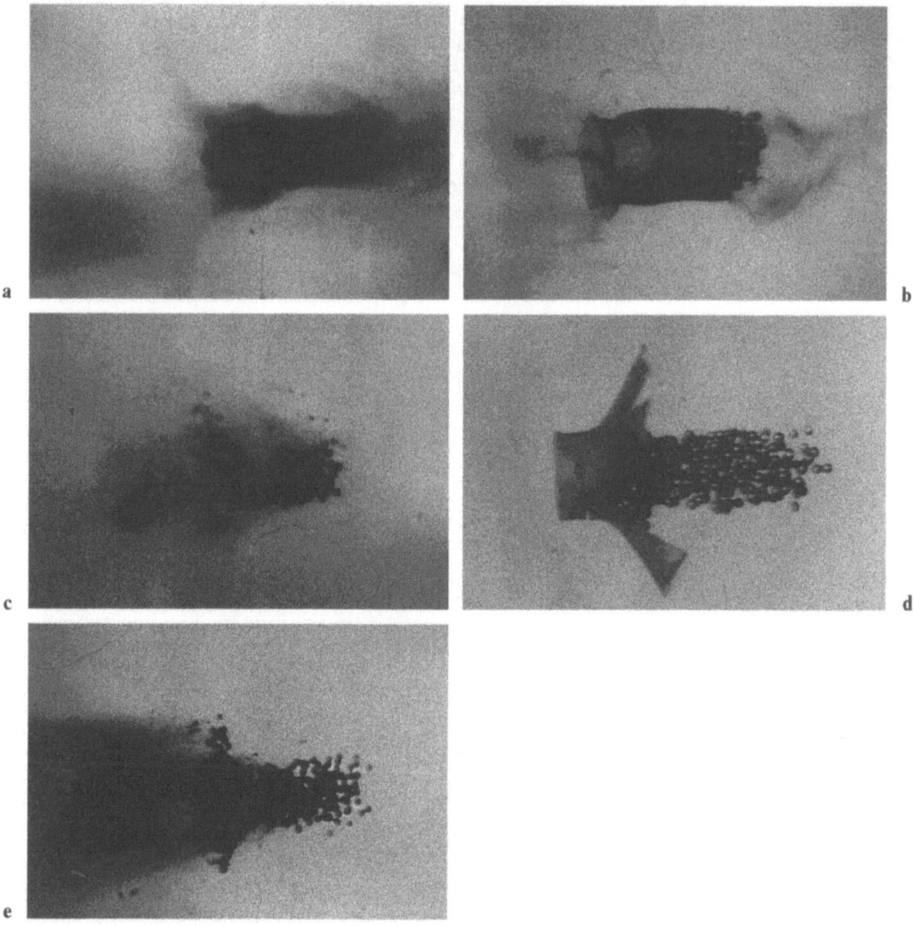

Fig. 57 a–e. Influence of shell construction on the shape of the shot column. Left column: shell with felt wad, Right column: shell with plastic cup, a) and b): shot column approx. 9 cm from the muzzle, c) and d): shot column approx. 60 cm from the muzzle, e) as c) photographed in transmitted light (for better visibility) (own photographs)

7.3 Shot Patterns—The Diameter of the Shot Pattern as a Function of the Range and Other Parameters

7.3.1 Theory

The shot pattern of shotguns at normal hunting ranges (approx. 20 to 40 m) has often been investigated. This field will therefore be discussed first.

The quantitative assessment of the shot pattern is made with a so-called 16-segment disk (Fig. 58). It has a diameter d of 75 cm, the inner one d/2 so that all the divisions have the same area. The pattern from a range of 20 m on such a disk

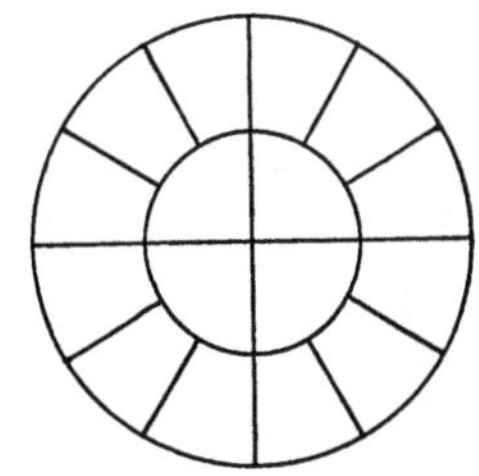

Fig. 58. Disk for assessing shot fired from a shotgun; outer diameter 75 cm, inner diameter 37.5 cm. The surface area of all the segments are equal

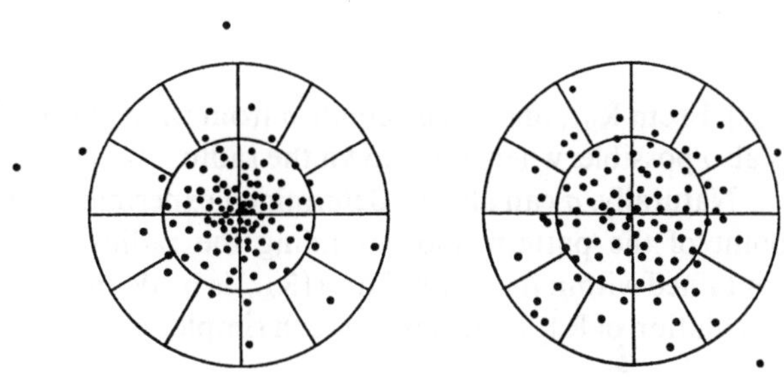

Fig. 59 a, b. Narrow and scattered shot at a range of 20 m. The figures show how the hits are distributed over the 16 segments of the disk from barrels with (**a**) strong and (**b**) weak constrictions

is shown in Fig. 59, from a barrel with a strong muzzle constriction in (a) and from a smaller constriction in (b). In the following the shot patterns will be seen from the theoretical view point.

Theoretically, a shot pattern should be uniform over its entire surface i.e. an equal distribution of the pellets (number of pellets per unit area) from the center to the edge. This requirement is not feasible in practice and the rotational—symmetrical distribution of the pellets follows, in reality, a two-dimensional, Gaussian distribution.

$$P(r) = 1 - \exp(-r^2/2 \cdot \sigma^2). \quad (1)$$

r = radius of the circle under observation (cm), σ = mean deviation (cm), P(r) = the fraction of the pellets in a circle of radius r(P = 1 for $r \gg \sigma$). The mean deviation σ can be clearly shown: in the circle with a radius σ lie 40% ($P\sigma) = 0.4$) of all the pellets, with $r = 2\sigma$ lie 87% ($P(2\sigma) = 0.87$), with $r = 3\sigma$ approx. 99% of the pellets lie within the circle (Table 11).

One can see that even when $r = 3\sigma$, 1% of all the pellets lie outside the circle. These "fliers" can be found at almost any distance from the center of the hit pattern. Their distribution can vary from shot to shot at the same range (Fig. 59 a + b, Fig. 60 c + d). The description of a hit pattern using the radius of the circle containing all the pellets would have a strongly varying size. Therefore, in general, the radius is given of *the* circle which contains 95% of all the hits (called

Table 11. Percentage of hits in a circle with a radius $r - \alpha \cdot \sigma$ with a known σ

α	$P(r) \cdot 100\%$
$\approx \frac{2}{3}$	20
1.0	40
2.0	87
2.15	90
2.45	95
3.0	99

R_{95}). From R_{95}, one can calculate σ from the equation $\sigma = R_{95}/2.45$. The reverse is also possible: when σ is known the radius R_{95} can be found from $R_{95} = 2.45 \cdot \sigma$.

Naturally, σ can also be determined experimentally by starting at the middle point of the pattern and extending the radius until 40% of all the hits are contained within the circle. This $r(R_{40})$ is by definition equal to σ (see above). The calculation of R_{95} is, however, much simpler in that one needs only to count from

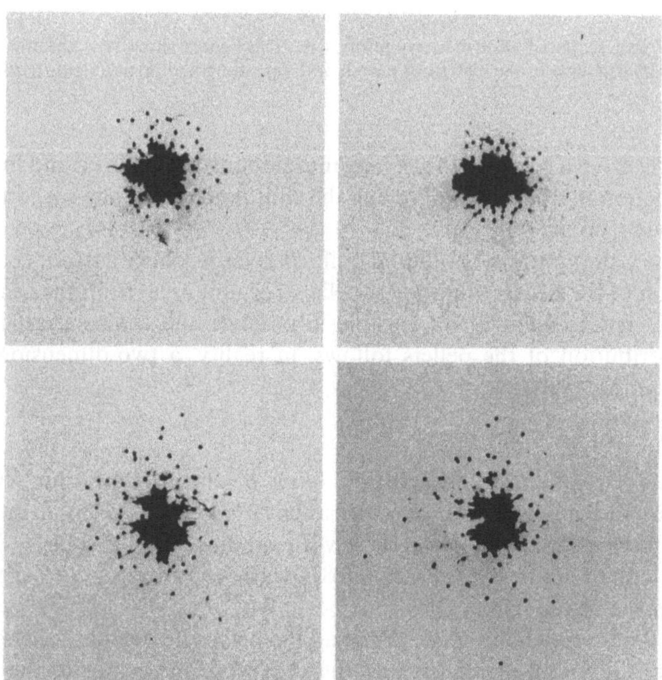

a b c d

Fig. 60 a–d. Variation in the hit pattern from shot to shot. All the shots are with $\frac{1}{2}$-choke. **a–b**) Top row: shot range = 1 m, **c–d**) Bottom row: shot range = 2 m. In spite of identical conditions, the hit pattern are not the same the periphery until one has eliminated 5% of the hits to obtain the 95% circle and with it the radius R_{95}.

It therefore follows that it is meaningless to quote the radius of a hit pattern unless the particular radius (R_{90}, R_{95}, etc.) is stated. It has already been said that the measurement of R_{100} (radius of the circle containing all the hits) makes no sense, because R_{100} varies greatly from shot to shot.

Measurement or calculation (e.g. from R_{95}) of σ is only valid for a given shot range s, for a given shell P, a given choke Ch and barrel length l. Therefore:

$$\sigma = f(s, P, Ch, l)$$

This means that with the same weapon and ammunition only σ is a function of the shot range:

$$\sigma = f(s)$$

Numerous investigations have been made in order to determine σ experimentally for various parameters P, Ch, l as a function of the shot range s.

These will be described in the next section.

7.3.2 Experimentally Determined Values of σ as a Function of the Shot Range and Other Parameter

In Fig. 61 (from Lampel-Seitz) the dependence of T_{90} as a function of the shot range and the type of choke is shown (the diameter of the hit pattern is given on the y-axis i.e. double the radius). The figure speaks for itself. σ is given by: $\sigma \approx R_{90}/2.3$.

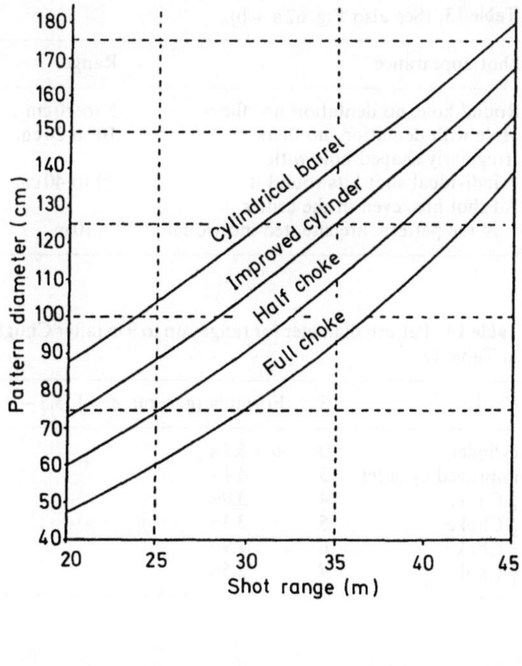

Fig. 61. Pattern diameter from conventional shells fired from a shotgun in relationship to range and type of bore (after Lampel-Seitz). ϕ_{95} is shown

Table 12. Pattern diameter ϕ as a function of shot range and the amount of choke (s in m, ϕ in cm) for ranges from 9 to 35 m (after Chugh)

Choke	Equation
cylindrical barrel	$\phi = 3.3 \cdot s + 20$
improved cylinder	$3.3 \cdot s + 5$
1/4 Choke	$3.3 \cdot s$
1/2 Choke	$s^2/17.4 + s/3.7 + 25$
3/4 Choke	$s^2/17.4 + s/3.7 + 20$
1/1 Choke	$s^2/17.4 + s/3.7 + 15$

According to the information in Table 11, the divisor 2.15 is used to calculate σ from R_{90}(for $R_{95} \rightarrow 2.45$). However, since the given radii are valid for 90–95% of the hits, R_{90} is divided by the mean value of 2.15 (R_{90}) and 2.45 (R_{95}).

Chugh (1970) gave empirical formulas for the pattern diameter as a function of shot range s and the degree of choke. These are given in Table 12 after being converted to the metric system.

According to the author, the deviations, with rare exceptions, are restricted to 2.5 cm (1 inch). Unfortunately, in the paper it is not specified which particular pattern diameter (R_{90}, R_{95}) is involved. Since the results of the fomula are in approximate agreement with those in Fig. 61, it may be assumed that ϕ_{90} to ϕ_{95} is the pattern diameter in question.

At ranges shorter than those in hunting (< 20 m), the rough rules given in Table 13 apply, yet again dependent on the degree of choke:

Table 13. (See also Fig. 62 a + b)

Shot appearance	Range
Round hole, no dentation nor fliers	5 to 10 cm
Hole with dentation, no fliers	10 to 20 cm
Irregularly shaped hole with individual shot hits round it	20 to 40 cm
All shot hits, even in the center of the pattern, are situated individually	$\geqslant$ 10 m

Table 14. Pattern diameter for ranges up to 9 m (after Chugh), otherwise as in Table 12

Choke	S	Formula (general: $\phi = [(5.3 - 0.4 \cdot S) \cdot s]$ [cm])
cylinder	0	$\phi = 5.3 \cdot s$
improved cylinder	3	$4.1 \cdot s$
$\frac{1}{4}$ Choke	4	$3.9 \cdot s$
$\frac{1}{2}$ Choke	5	$3.3 \cdot s$
$\frac{3}{4}$ Choke	6	$2.9 \cdot s$
$\frac{1}{1}$ Choke	7	$2.5 \cdot s$

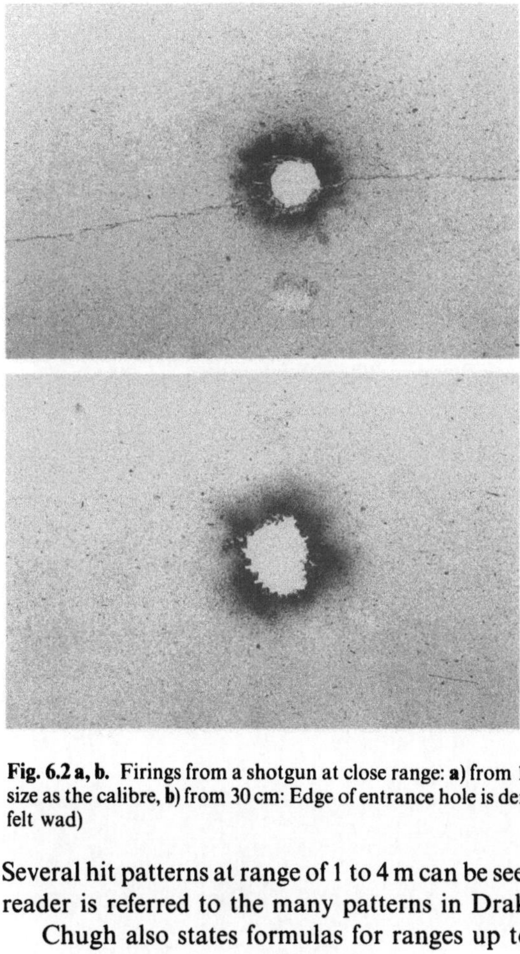

a

b

Fig. 6.2 a, b. Firings from a shotgun at close range: **a)** from 10 cm: Entrance hole round and the same size as the calibre, **b)** from 30 cm: Edge of entrance hole is dentated, several fliers are visible (shell with felt wad)

Several hit patterns at range of 1 to 4 m can be seen in Fig. 63 a–h. In addition, the reader is referred to the many patterns in Drake (1961/62).

Chugh also states formulas for ranges up to 9 m:

$$\phi = (5.3 - 0.4 \cdot S) \cdot s \, [\text{cm}], \qquad (1)$$

whereby the values of S are dependent on the degree of choke (see Table 14).

Heany and Rowe (1983) shot at targets using a 12 gauge shotgun, with cylindrical bore, Remington shells with 3.8 mm pellets at ranges from 3.9 to 15.2 m in 1.5 m steps. They used the R_{100} circle for their calculations and were of the opinion that fliers would not significantly affect the results. The construction of the linear regression for the description of measurements obtained gave

$$\phi = -6.8 + 5.7 \cdot s \, [\text{cm}] \quad \text{with} \quad r = 0.995 \qquad (2)$$

for the range of s: 3.9...15.2 m.

More theoretical investigations were made recently by Bhattacharyya et al. (1989) and Boyer et al. (1989).

When one takes into consideration that Chugh uses ϕ_{90} to ϕ_{95} in his equations and Heaney and Rowe, in contrast, ϕ_{100}, the equations are in

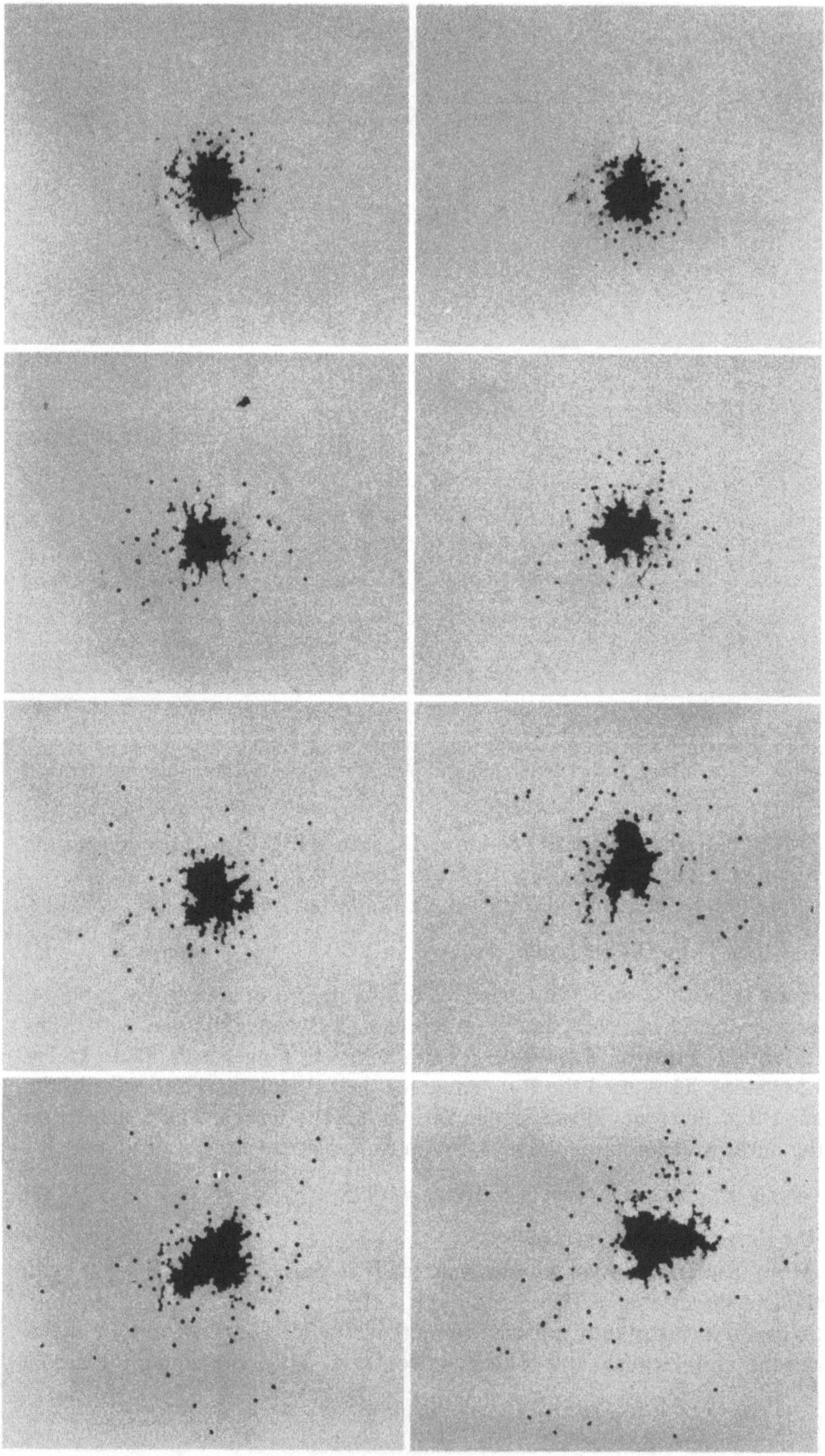

agreement with each other even to the constant − 6.8. Both equations give sufficiently useful values that within the framework of deviation are in agreement. The authors give as the standard deviation for the equations ± 0.14 m. This value may be mathematically correct but in my experience it is not achieved by a long way in practice.

Speak et al. (1985) investigated the properties of the hit pattern for shot shells in the calibers .22, .38, and .44 at shot ranges between 0.3 and 6.1 m. A linear regression was constructed with the data obtained; however, neither R nor ϕ were used to describe the hit pattern but rather the expression $\sqrt{A}$. A is defined as being the area of the smallest rectangle which just contains the complete hit pattern. This quantity $\sqrt{A}$ is theoretically $\sqrt{(\frac{1}{4}\pi)}$ larger than ϕ_{100} as is easily derived from the size relationship of the rectangle (square) to the equivalent circle. With an oval distribution (which is quite possible with a hit pattern with unsymmetrical flyers) the quantity $\sqrt{A}$ is dependent on how the rectangle has to be fitted to the hit pattern.

The authors use a linear regression of the form:

$$\sqrt{A} = a + b \cdot s \,[\mathrm{cm}]. \tag{3}$$

The value of *a* varies with the shells measured by between − 0.8 and 2.8 cm, the value of *b* between 8.5 and 13.5 cm/m. So, for example, for a .22 l.r. (CCI) shell

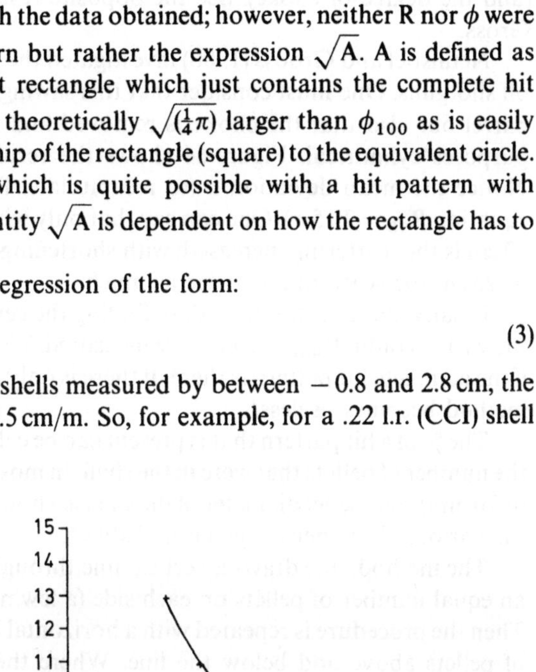

Fig. 64. Scattering of pellets as a function of barrel length and shot range s for 3.5 mm pellets. The y-axis: $\sigma_{1.s}$ ($s = 2 \cdot R_{39}$, see e.g. Sellier 1982) (from Nennstiel and Grooss 1980)

Fig. 63. Hit pattern in relationship to range and type of choke. Left column: full choke, right column: half choke. From top to bottom ranges of 1, 2, 3 and 4 m.

from a Smith and Wesson revolver Mod.34-1 (1 = 5 cm) the equation is as follows:

$$\sqrt{A} = 2.8 + 10.2 \cdot s\,[cm] \qquad (r = 0.996). \tag{3a}$$

With the same shell fired from a rifle, in spite of the considerably longer barrel length (1 = 61 cm), the same dependence as in Eq. (3a) is found. This may be because this shell has a plastic cup which reduces the effect of the barrel length (and the degree of choke). See the opposite results obtained by Nennstiel and Gross.

Nennstiel and Grooss (1980) investigated the scattering of pellets from sawn-off shotguns. One must consider that this sawing off produces not just a shorter barrel but also that the choke is removed and the barrel is to all intents and purposes cylindrical. Figure 64 shows the results obtained by these authors. Notice the mean deviation σ (to recapitulate: R_{90} can be calculated from the equation $R_{90} = 2.15 \cdot \sigma$). One can say that only when the barrel length is less than 40 cm is the scattering increased; with shortening of the barrel down to about 20 to 25 cm the scattering is twice as much.

Finally, there is still the task of finding the center of a hit pattern from which the various radii (R_{90}, R_{95}, etc.) are measured. The method is simple and accurate if the complete hit pattern is there. If there is only a part of it then, in principle, the method becomes useless.

The % of a hit pattern that is present can be calculated with a comparison with the number of pellets that were in the shell. In most cases the caliber, the length (65 or 70 mm), and pellet diameter of the shell is known. The total number of pellets in the various shot shells is given in Table 5.

The method: one draws a vertical line through the hit pattern so that there is an equal number of pellets on each side (a few more on one side is not critical). Then the procedure is repeated with a horizontal line so that there equal numbers of pellets above and below the line. Where the two lines cross, is with great accuracy, the center of the hit pattern and the starting point for measuring radii. It is easy to recognize that in each quadrant formed by the vertical and horizontal lines lie 25% of the hits.

7.4 Practical Procedures of Shot Range Determination for Shotguns

The determination of the range is simple and relatively accurate when the complete hit pattern is available for examination. This is the case when a person has been shot from a range of 10 m or less and the aiming point was the center of the body. With such a shot, all the pellets (excepting possible fliers) are to be found on or in the body. Another possibility is that the target aimed at is standing in front of a large flat surface (side of a house, a wall, or a barn which is good when built of wood as it is then easy to find the pellets). It is a decisive factor if the complete hit pattern can be reconstructed or not.

Even when all these ideal relationships are present and the complete hit pattern is available, no great accuracy can be expected. Anyone who has experience and has had to assess many hit patterns knows about their fickleness. No hit pattern resembles another, even from the same weapon with the same type of shell at the same range. In every case, enough (!) test shots must be made with the actual weapon and the same ammunition so that the range can be determined by comparison of the hit pattern in the actual case with those from the test shots.

It is difficult, if not impossible, to determine the shot range when only a section of the hit pattern is available. This occurs when a person is shot from long range (> 15 to 20 m) because the area of the hit pattern is greater than that of the target surface of the person. It also happens when the column of shot misses the target which is then only struck by fliers.

As already described in Sect 7.3.1, the density of pellets over the hit pattern is not evenly distributed but becomes less as it approaches the edge (Gaussian distribution). A particular density in the outer part of the hit pattern from e.g. 10 m range can be identical with the density from the center of a hit pattern from a range of 30 m. This ambiguity increases as the area under investigation becomes smaller.

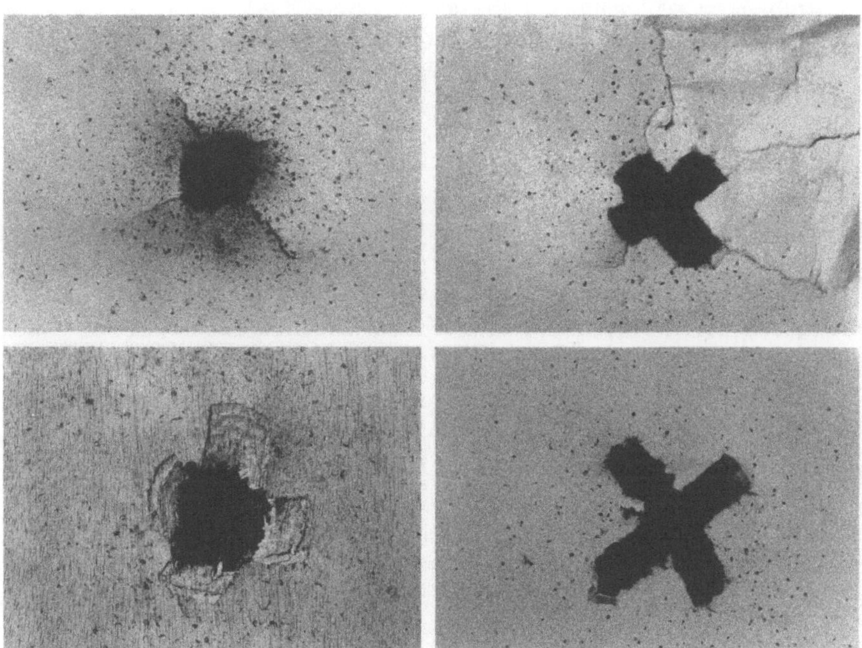

a b c d

Fig. 65 a–d. Effect from the opened out plastic cup at intermediate range. Shell 12/70 Remington: **a)** Range 30 cm: cup not yet opened out. **b)** Range 50 cm: cup almost completely opened out. **c)** Range 50 cms as **b)**: effect on wood; notice the deep impression left by the light plastic wings. **d)** Range 60 cm: completely opened out. At ranges of 70 cm and more only random variations are observed: the wings have been bent back completely. Notice the powder tattooing (from Sellier 1982)

An extreme example would be hits from two pellets in an area the size of one's palm of the hand—this could be 2 fliers from a 2 m hit pattern or could be from the central part of a 40 m hit pattern! Here, one can only consider the results very carefully and critically assess them.

Shots from ranges of less than 10 m can be determined relatively accurately because in most cases the complete hit pattern is visible due to the small pattern radius and its changing characteristically with increasing range (Table 7.3.2:2). With modern ammunition which has a plastic cup, the degree of opening out of this cup can also used to determine range (Janssen and Kistner 1968). See

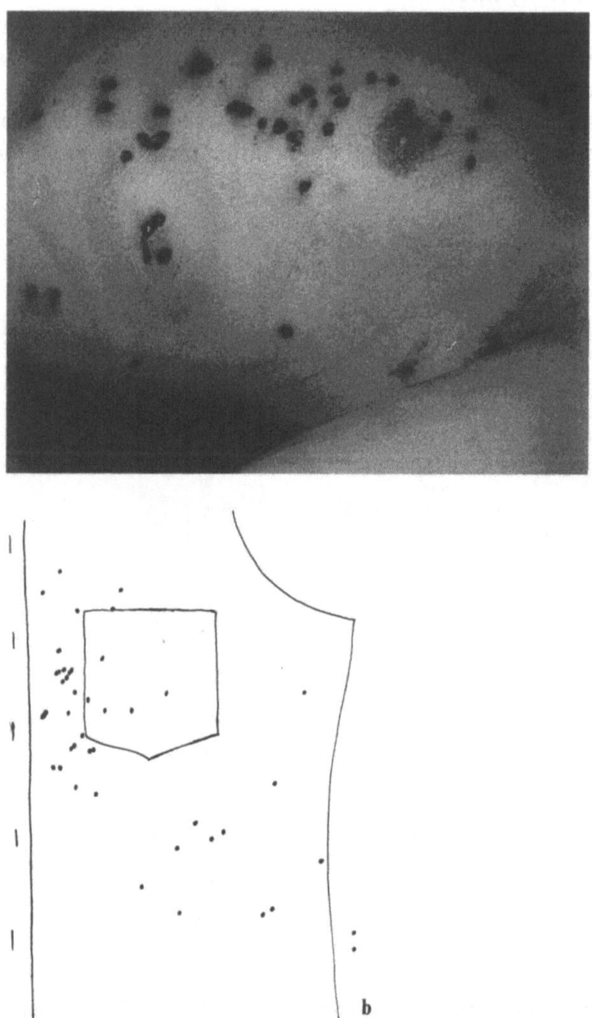

a

b

Fig. 66 a, b. "Screening" of a hit pattern by an obstacle (wall); for further details see text. **a**) Hit pattern on body, **b**) Hit pattern on the shirt that was being worn.

Fig. 57 d where the cup is almost completely opened (range = 0.6 m). Naturally, the degree of opening out is not just a factor of range (Fig. 65 a–d) but also dependent on the construction, thus varying from manufacturer to manufacturer making test shots necessary.

Errors in assessing hit patterns, and therefore ranges, can occur when an intermediate target acts as a screen or a further source of scattering.

Screening off of a part of the hit pattern is often a sign that the victim was seeking shelter from the shot by standing close behind a wall.

I once had to investigate such a case. I was given the clothing and a picture showing the injuries of the deceased. Without any further information about the circumstances, I was asked to determine the range (Fig. 66). The centerline of the front of the body formed the right-hand margin of the hit pattern, moving left the density of pellets decreased partly due to the oblique impact on the side of the chest wall (projection distortion, see below). A distribution where the density is greater at periphery than at the center does not exist. (This statement applies to the average; for small regions of the pattern it is possible that by chance the density is greater nearer the edge than in the center). For this reason it was impossible for this (relatively sharp) boundary on the middle of the body to be the edge of the hit pattern. From the distribution which had been determined it was clear that there must have been an intermediate target with a vertical edge. The simplest explanation was that the victim had sought protection behind a wall and so it proved to be from examination at the scene of crime.

When the pellets strike an inclined surface, the density appears less. When the axis of rotation of the target's surface is vertical to the (assumed) horizontal firing direction, the density in the horizontal direction decreases whereas in the vertical direction it remains the same. If the surface is randomly situated to the direction of firing, the whole hit pattern is distorted. This can be taken into consideration when the position of the target in regard to the direction of firing is known.

The distribution of the pellets on the target can be changed by an intermediate target. The scattering increases in that the density decreases. Intermediate targets can consist of bushes, thin walls or sheets of glass. Thick or hard intermediate targets (depending on the diameter of the pellets) do not come into contention because they stop the pellets completely. Shooting through bushes not only increases the scattering but also absorbs some of the pellets (naturally more with a small diameter (2.5 mm) than with a larger one (4 mm)). This must obviously be considered when assessing the hit pattern (apparent reduction of hit density caused by retained pellets in branches of bushes).

In one case, a guard fired from a car through the windshield (6 mm thick) with a shotgun and a 12/70 shell (3.5 mm ϕ pellets) at a person 10 m away. Test shots showed that the scattering under these conditions was twice the normal.

The increased scattering due to an intermediate target is dependent on various factors, in the main, on the pellet size, the characteristics of the intermediate target, and if the charge of shot strikes as a whole or each pellet separately. The distribution of the pellets visible in the depth of the body or part of the body (X-ray picture) can exhibit an increased scattering caused by tissues. This could lead to an erroneous evaluation because the apparent reduction in density would make the shot range determination too large. It is therefore advisable that the hit pattern from the X-ray plate should not be used for range determination but rather from the skin or the clothing worn because then scattering by the tissues of the body is then eliminated.

This possibility does not exist for bodies which have been burned since the skin injuries and the clothing are lost. In such cases, the question is raised as to whether the hit pattern in the body (obviously from the X-ray picture) can be used to reconstruct the undistorted pattern and therefore the true range.

Breitenecker and Senior (1967) carried out numerous experiments on this problem and came to the following conclusions: When the pellets all strike the intermediate target together, they influence each other and scatter more than without an intermediate target (here: skin and tissues).

However, when all the pellets strike singly and remain buried in the body, the scattering shown in the X-ray picture is practically unaffected by the intermediate target (skin and tissue). The range can therefore be determined from the X-ray picture after the enlargement is allowed for.

8 References and Further Reading

Alfonsi A, Calatri S, Cerioni E, Luchi P (1984) Shooting distance estimation for shots fired by a shotgun loaded with buckshot cartridges. Forens Sci Internat 25: 83

Andrasko J, Maehly AC (1977) Detection of gunshot residues on hands by scanning electron microscopy. J Forens Sci 22: 279

Aragona F, Crino C, Bonavita V (1977) Il cosidetto "tatuaggio" da polveri infumi nelle ferite d'arma da fuoco recenti. Zacchia 13: 68

Baima Ballone PL, De Bernardi A, Martini M (1967) Gli aspetti del tatuaggio prodotte dale moderne polveri infumi nei colpi esplosi con armi corte. Arch Soc lombarda Med leg 3: 380

Baisane SO, Chincholkar VS, Mattoo BN (1979) Spectrographic determination of barium in biological material. Forens Sci Internat 12: 127

Bakonyi E, Farago E, Tomcsnyi R (1970) Lassen die unterschiedlichen CO-Hämoglobin-Konzentrationen im Bereich von Ein- und Ausschuß Aussagen über die Schußrichtung und -entfernung zu? Arch Krimin 145: 35

Barnes FC, Helson RA (1974) An empirical study of gunpowder residue patterns. J Forens Sci 19: 448

Barsegyantz LO, Vereshchaka MF (1981) Firearm injuries to the hairs. (Russian with English summary). Sudebnomed esp 14, Nr 4: 34

Barz J (1972) Die Bestimmung der Schußentfernung aus unverbrannten Pulverteilchen. Beitr gerichtl Med 29: 327

Barz J, Löhle U, Eitner F (1975) Zum Nachweis des absoluten Nahschusses. Vergleichende Untersuchungen mittels (Lichtmikroshop, flammenloser Atomarbsorption und Röntgenspektrometrie. Beitr gerichtl Med 23: 273

Barz J (1971) Die Bestimmung der Schußentfernung aus unverbrannten Pulverteilchen. ID Heidelberg

Basu S (1982) Formation of gunshot residues. J Forens Sci 27: 72

Baumgärtner F, Schöntag A (1962) Aktivierungsanalyse in der Kriminaltechnik. Kerntechnik 4: 51

Baumgärtner F, Stärk H, Schöntag A (1963) Aktivierungsanalytische Antimonbestimmung im Nanogrammbereich zur Ermittlung von Schußentfernungen. Z Analyt Chem 197: 424

Belov AP (1954) Über die Temperatureinwirkung der Pulvergase auf die Gewebe der Kleidung b. Schuß aus d. Pistolette. Probl d gerichtsmed Expertise S 7 (Mosk.)

Berg S (1959) Veränderungen der Textiloberfläche bei Nahschüssen. Arch Krimin 124: 5–8/17

Berg S, Kijewski H (1982) Vermeintliche und wirkliche Nahschußzeichen am KnocheneinschuBloch des Schädels. Z Rechtsmed 88: 103

Bergman P, Springer E (1987) Bullet Hole Identification Kit: Case Report. J Forens Sci 32: 802

Bhattacharyya C, Sengupta PK (1989) Shotgun Pellet Dispersion in a Maxwellian Model. Forens Sci Internat 41: 205

Böhm E (1972) Untersuchungen an Kopfhaaren im Nahschußbereich mit dem Rasterelektronenmikroskop. Arch Krimin 149: 65

Böhm E, Klingele H (1972) Morphologische Untersuchungen an Pulverschmauchpartikeln verschiedener Munitionsarten. Arch Krimin 150: 31

Böhm E, Burchard WG, Klar JJ, Wehr K (1974) Morphologie und Elementverteilung an Ein- und Ausschuß. Beitr gerichtl Med 32: 242

Böhm E, Schreiber HP und Suchenwirth H (1969) Röntgenologische Untersuchungen zur Schußentfernungsbestimmung. Dtsch Z gerichtl Med 65: 112

Bohne G (1954) Nah- oder Fernschuß. Kriminalistik 8: 126

Bohne G (1954) Nah- oder Fernschuß. Umschau Heft 17: 521

Bonnet EFP, Pedace EA (1971) Signe de Benassi; ratification histopathologique de son importance médico-légale. Méd lég Dommage corp 4: 46

Bonorden T, Schewe G, Lins G (1973) Schrotschußverletzungen mit moderner Plastikbecher Munition. Arch Krimin 152: 20

Booker JL, Schroeder DD, Propp JH (1984) A note on the variability of barium and antimony levels in cartridge primers and its implication for gunshot residue identification. J Forens Sci Soc 24: No 2: 81

Bosch K, Gruber H (1975) Ein Beitrag zur chemischen Darstellung von Nahschußpuren. Kriminalistik 29: 315

Bosen SF (1976) A rapid microtechnique for the detection of trace metals from gunshot residues. J Forens Sci 21: 163

Botta I, Vicol I (1964) Determining distance of shooting by means of alpha-naphthylamine (Rumanian). Probl Med judic crim (Budapest) 2: 177

Boyer DA, Marshall LD, Trzicak LM, Rowe WF (1989) Experimental evaluation of the distribution of the pellets in shotgun pellet patterns. Forens Sci Internat 42: 51

Breitenecker R, Senior W (1967) I. An experimental study on the influence of intermediate targets. J Forens Sci 12: 193

Brihaye Cl, Machiroux R, Gillain G (1982) Gunpowder residues detection by anodic stripping voltammetry. Forens Sci Int 20: 269

Brüning A, Schnetka M (1937) Über die chemische Untersuchung und die Beurteilung von Einschüssen. Arch Krimin 101: 81

Burdelski M (1968) Erweiterung des Schußentfernungsbereiches durch Anreicherung der Schmauchelemente. ID Bonn

Bürger H, Neuninger (1970) Röntgenfluoreszenz- und emissionsspektralanalytische Schußentfernungsbestimmung. Arch Krimin 145: 11

Burkert, Schöntag A (1970) Eine schwerwiegende Fehlermöglichkeit bei der spektrografischen Bestimmung der Schußentfernung an der Haut oder an Textilgeweben. Arch Krimin 145: 17

Burnett B (1989) The form of gunshot residue is modified by target impact. J Forens Sci 34: 808

Calabuig JAG, Carreras LF (1976) Tatouage radio-opaque dans les blessures par projectiles d'armes à feu. Acta Méd lég et soc 27: 133

Candela RG, Colonna M, Strada L (1976) Identificazione dei residui dell'esplosione di colpi d'arma da fuoco corta mediante polarografia di stripaggio anodico. Zacchia 51:44

Chiarotti M, Fucci N, Luccini P, Carnevale A (1987) Thin-layer and high performance liquid chromatographic analyses of organic gunshot residues. Jpn J Legal Med 41: 65

Chugh OP (1970) Empirische Untersuchung über die Streuung der Schrote beim Schuß. Arch Krimin 146: 109, 168 (engl)

Chugh OP (1986) Mathematical analysis of dispersion of pellets fired from a shotgun. Forens Sci Intern 32: 93

Chupikov AS (1984) The examination of gunshot traces in case of close shooting from Dragunov sniper's rifle (russ.). Subednomed eksp (Mosk) Nr.1: 25

Cimino AM, Debney JE (1968) Distance of shot from discharge residues of rifled weapons. J Forens Sci Soc 8: 8

Coleman RF (1966) The application of neutron activation analysis to forensic science. J Forens Sci Soc 6: 19

Cone RD (1973) Detection of barium, antimony and lead in gunshot residue by flameless atomic absorption spectrophotometry. Police Weapons Center Bull 6: 4

Copeland AR (1981) An improved method of gunshot wound examination. J Forens Sci 26: 552

Cornelis R, Timpermann J (1974) Gunfiring detection method based on Sb, Ba, Pb and Hg deposits on hands. Evaluation of the Credibil. Med Sci Law 14: 98

Cranz C Lehrbuch der Ballistik. Springer Bd.I, 5.Aufl., Bd.II 1926, Bd.III 1927

Dahl DB, Cayton JC, Lott PF (1987) Gunshot residue analysis: an application study. Microchem J 35: 360
Deinert W, Leszczynski Ch (1986) Examinations to determine close-range firing distances using a process control computer. Forens Sci Intern 31: 41
Dimaio VJM, Zumwalt RE (1977) Rifle wounds from high velocity, center-fire hunting ammunition. J Forens Sci 22: 132
Dimaio VJM, Petty CS (1978) The effectiveness of the sawed-off shotgun. Forens Sci Gazette 9, Nr 2: 5
Dimaio VJM, Petty CS, Stone Jr. CS (1976) An experimental study of powder tattooing of the skin. J Forens Sci 21: 367
Dimaio VJM, Spitz W (1970) Injury by birdshot. J Forens Sci 15: 396
Dowing GP, Dickinson JAH, Cooke CT (1988) Shotcup petal abrasions in close ranges. 410 caliber shotgun injuries. J Forens Sci 33: 260
Drake V (1961/62) Shotgun Ballistics-1 J Forens Sci 2: 85
Drake, V (1962) Shotgun Ballistics-2 J Forens Sci 3: 22
Eidlin LM (1961) Glycerine test, a new method of detecting gem powder and its remnants at the zone of gunshot injury. Sudebnomed eksp (Moskau) 4, Nr 4: 22 (Russian)
Eidlin LM (1961) One of the causes of peculiar course and outcome of gun-shot wounds. (in Russian). Sudebnomed eksp 4: 22
Eigendorf H, Schmechta G (1979) Grenzen der Schußentfernungsbestimmung bei Nahschüssen mit Druckluftwaffen. Kriminal Forens Wiss Nr. 35: 15
Eigendorf H-G, Schmechta H (1982) Voraussetzungen und praktischer Wert der Schußentfernungsbestimmung mit quantitativen Methoden. Kriminalistik u Forens Wissenschft : 5
Eigendorf H-G, Schmechta H (1979) Untersuchungen über die Dichteverteilung der Schmauchelemente Pb, Sb, u. Cu bei Schüssen aus der Pistole MAKAROW 9 mm. 1. Mitt. :Radiale Verteilung. Kriminalistik u Forens Wissenschft : 23
Eigendorf H-G, Schmechta H (1979) Untersuchungen über die Dichteverteilung der Schmauchelemente Pb, Sb, u. Cu bei Schüssen aus der Pistole MAKAROW 9 mm. 1. Mitt.: Radiale Verteilung. Kriminalistik u Forens Wissenschft :17
Eigendorf H-G, Schmechta H (1980) Asymetrisches Schmauchbild nach Veränderung der Waffenmündung der Maschinenpistole "Kalaschnikow". Kriminalistik u Forens Wissenschft : 47
Eigendorf HG, Schmechta H (1982) Verteilung der Schmauchelemente auf der beschossenen Oberfläche nach Veränderung der Waffenmündung der Maschinenpistole "Kalaschnikow". Kriminal Forens Wiss Nr 46: 77
Elbel H (1939) Schußwinkel und Schmauchbild. Dtsch Z gerichtl Med 32: 165
Even H, Bergman P, Springer E, Klein A (1988) The effects of water-soaking on firing distance estimations. J Forens Sci 33: 319
Fackler ML, Woychesin SD, Malinowski JA et al. (1987) Determination of shooting distance from deformation of the recovered bullet. J Forens Sci 32: 1131
Fallani M (1953) Di un nuovo reattivo il riconoscimento dei nitrici e della sua pratica applicabilita in med. leg. Minerva medicoleg 73: 101
Fann Ch F, Ritter WA, Watts RH, Rowe WF (1986) Regression analysis applied to shotgun range-of-fire estimations: results of a blind study. J Forens Sci 31: 840
Fendrich K (1966) Die Beeinflussung der Schußentfernungsbestimmung durch Wind. ID Bonn
Fritz E (1937) Die Erkennung vom Nahschuß in der behaarten Kopfhaut. Dtsch Z gerichtl Med 28: 215
Fritz E (1942) Eigenartige Schmauchbilder bei absolutem Nahschuß. Dtsch Z gerichtl Med 35: 13
Garsche R (1935) Die Stanzmarke, ein Zeichen des absoluten Nahschusses. Arch Krimin 97: 120
Gliko LI (1968) Über die Reproduzierbarkeit der Nahschußzeichen aus der Pistole "PM" und d. automat. Pistol "AK". Sudebnomed eksp (Moskau) : 10
Goleb JA, Midkiff Jr CR (1975) The determination of barium and antimony in gunshot residue by flameless AAS using a tantal strip atomizer. Appl Spectroscopy 29: 44
Gruber H (1974) Ein Verfahren zur Bestimmung der Schußentfernung durch elektrographische Untersuchung an Textilien. Beitr gerichtl Med 32: 240
Hanke CW, Conner AC, Probst EL, Fondak AA (1987) Blast tattoos resulting from black powder fire-arms. J Am Acad Dermatol 17: 819
Harrison HC, Gilroy R (1959) Firearms discharge residues. J Forens Sci 4: 184

Hausbrandt F (1944) Experimentelle Studien zur Entstehungsmechanik und Morphologie einiger Nahschußzeichen. Dtsch Z gerichtl Med 38: 45

Heaney KD, Rowe WF (1983) The application of linear regression to range-of-fire estimates based on the spread of shotgun pellet pattern. J Forens Sci 28: 433

Helbig W (1975) Einfaches und rasches Verfahren zur Sichtbarmachen von Pulverrückständen. Kriminalistik Forens Wiss 15: 63

Hellmiss G, Lichtenberg W, Weiss M (1987) Investigation of gunshot residues by means of Auger electron spectroscopy. J Forens Sci 32: 747

Heuschkel HJ (1979) Nachweis des absoluten Nahschusses trotz operativer Entfern. d. Hautstücks mit der Einschußöffnung. Z ärztl. Fortbid 73: 567

Hochmeister M, Dirnhofer R (1989) "Krähenfußbildungen" im Schmauchbild bei Tötung durch Stirnschuß. Z Rechtsmed 102: 545

Hoffmann CM (1973) Neutron activation analysis for the detection of firearm discharge residue collected on cotton sw. JAOAC 56: 1388

Hoffmann R (1985) Über die Anwendung des Folienabdruckverfahrens—ein Verfahren zum Nachweis der Flächenverteilung von chemischen Elementen in Schußresiduen. Arch Krimin 175: 21

Hoffmann R (1987) Geisterschmauch—eine Fehlermöglichkeit bei der Schußspurenuntersuchung. Arch f Krim 179: 13

Holsten K (1936) Zur Frage der Schußentfernungsbestimmung bei Verwendung von Sinoxidmunition. Dtsch Z gerichtl Med 26: 389

Holsten K (1937) Untersuchung über die Nachweismethoden und über die Verteilung des Bleis im Schußfeld. Dtsch Z gerichtl Med 28: 205

Huber O (1968) Eine einfache Methode zum qualitativen Nachweis von Blei bei Nahschuß. Arch Krimin 141: 45

Hunger H, Zerndt B, Leopold D (1964) Zur forensischen Beurteilung von Nahschüssen und Feststellung ihrer Priorität. Arch Kriminol 134: 69

Hunt AC, Kon VM (1962) The pattern of injury from humane killers. Med Sci Law 2: 197

Janezic-Jelacic O (1953) L'effet thermique produit sur les plaies par armes à feu. Ann Méd lég etc 33: 273

Janssen W (1966) Nahschußzeichen auf Kunstfaserstoffen unter besonderer Berücksichtigung der Oberflächenstruktur. Dtsch Z gerlichtl Med 58: 112

Janssen W, Kistner D (1968) Nahschußverletzungen durch neuartige Schrotmunition unter besond. Berücksicht. der Schrotstreuung. Arch Kriminol 141: 3

Janssen W, Stieger W (1964) Verletzungen durch Bolzenschuß-Apparate unter besonderer Berücksichtigung der Spurenmerkmale. Arch Kriminol 134: 26

Jauhari M. Chatterjee SM, Ghosh PK (1972) Statistical treatment of pellet dispersion data for estimating range of firing. J Forens Sci 16: 141

Jones PF, Nesbitt RS (1975) A photoluminescence technique for detection of gunshot residues. Forens Sci 20: 231

Kamaka SK (1970) Sawed-off shotguns and shotgun patterns. AFTE J 26

Kapusz N, Nagy L (1967) Schußspuren an Textilien aus Kunststoffasern. Arch Krimin 140: 149

Kijewski H (1977) Die Nachweisbarkeit von Schmauchelementen nach Einwirkung von Wasser und Feuer. Arch Krimin 159: 23

Kijewski H (1981) Die Bedeutung der Schußentfernungsbestimmung für die Rekonstruktion von Schußwaffendelikten. Kriminalistik 35: 369

Kijewski H (1974) Neue Aspekte in der Schußentfernungsbestimmung durch Anwendung der flammenlosen Atom-Absorptions-Spektralphotometrie. Beitr gerichtl Med 32: 238

Kijewski H, Bock G (1983) Veränderungen der Sb-und Pb-Konz.im Abstreifring bei Schüssen mit Langwaffen aus Entfern. bis 400 m. Beitr gerichtl Med 41: 383

Kijewski H, Lange J (1974) Möglichkeiten und Grenzen einer Schußentfernungsbestimmung aus dem Schmutzring. Arch Krimin 154: 179

Kijewski H, Möhle M (1976) Der mikroanalytische Nachweis von Kupfer mit Dithiooxamid als Grundlage einer neuen Methode der Schußentfernungsbestimmung. Z Rechtsmed 77: 121

Kijewski H, Jäkel M (1986) Die Waffenölmenge am Ziel als Funktion der Schußreihenfolge und der Schußentfernung. Z Rechtsmed 97: 111

Kinard WD, Midkiff Jr CR (1978) The application of oxygen plasma ashing to gunshot residue Analysis. J Forens Sci 23: 368

Klages U, Münstedt H, Janssen W (1971) Nahschußzeichen bei Handfeuerwaffen mit manipulierter Mündung. Arch Krimin 148: 146
Koch, OG, Koch-Dedic GA (1964) Handbuch der Spurenanalyse, Springer-Verlag
Kohlschütter H (1937) Zur Frage der Schußentfernungsbestimmung für automatische Repetierpistolen bei Verwendung von Sinoxydmunition. ID Göttingen
Koons RD, Havekost DG, Peters Ch A (1988) Determination of Ba in gunshot residue collection swabs using inductively coupled plasma-atomic ES. J Forens Sci 33: 35
Krishnan SS (1967) Determination of gunshot firing distances and identification of bullet holes by NAA. J Forens Sci 12: 112
Krishnan SS (1967) Firing distance determination by neutron activation analysis. J Forens Sci 12: 471
Krishnan SS (1973) Trace element analysis by atomic absorption spectrometry and NAA in the investigation of shooting cases. J Canad Soc Forens Sci 6: 55
Krishnan SS (1974) Firing distance determination by atomic absorption spectrophotometry. J Forens Sci 19: 351
Krishnan SS, Nichol RC (1968) Identification of bullet holes by neutron activation analysis and autoradiography. J Forens Sci 13: 519
Krishnan SS, Gillespie KA, Anderson EJ (1971) Rapid detection of firearm discharge residues by atomic absorption and neutron activation analysis. J Forens Sci 16: 144
Kübler J (1941) Über Veränderungen an Nahschußspuren auf menschlicher Haut unter der Einwirk. äußerer Einflüsse. ID Zürich
Kupov IY (1968) Schußentfernung und Waffenbestimmung mit einer modifizierten Methode des Abdrucks (in Russian). Sudebnomed eksp (Moskau) 11; Nr 4: 17
Kustanovic SD (1954) Die Temperaturwirkung eines absoluten Nahschusses mit Patronen mit rauchlosem Pulver auf dem Gebiet der Einschußöffnung. Probl d gerichtsmed Expert Staatsverl Mosk
Kutterer RE (1959) Ballistik. Vieweg & Sohn, Braunschweig
Kyassi AI, Hawkins ES (1938) Tatooing by cordite discharged from rifles. Verh 1 Intern Kongr gerichtl Med 591
Lampel W, Mahrholdt R (1975) Waffen-Lexikon. Bayrischer Landwirtschaftsverlag BLV, München
Lampel W, Seitz G (1983) Jagdballistik. Neumann-Neudamm, Melsungen 3 edn
Lautenhach L (1969) Über das Auftreten und das Ausmaß von Schmauchspuren an der Innenseite von Kleidungsstücken beim Nahschuß. Beitr gerichtl Med 26: 108
Leffers B, Jeanty D (1982) Handgun pellet ammunition ("snake shot") wounds: report of three cases. J Forens Sci 27: 433
Lekstrom JA, Koons BS, Koons RD (1986) Copper and Nickel Detection on Gunshot Targets by Dithiooxamide Test. J Forens Sci 31: 1283
Lenoir L, Debarge A (1988) Intérêts et limites de l'étude histologique et radiologique des plaies cutanées par balle de 22 LR. J Méd lég 31: 443
Leszczynski Ch (1968) Auswertung von Schußspuren mit Hilfe eines digitalen Prozeßrechners. Vortr Jahrestagung Rechtsmed Innsbruck Okt
Leszczynski Ch (1959) Die Bestimmung der Schußentfernung. Kriminalistik 13: 377
Lichtenberg W (1985) Schußspurenuntersuchung in Verbindung mit Munitionsarten mit "bleifreiem" Pulverschmauch. Beitr gerichtl Med 43: 293
Lichtenberg W (1983) Untersuchung des Pulverschmauchs einer Munition mit bleifreien Anzündsäzen. Kriminalistik 37: 590
Lichtenberg W (1988) Die Munition wurde "bleifrei". Der Pulverschmauch neuerer Munitionsarten m. bleifreien Anzündsäzen. Kriminalistik 42: 73
Liebhardt E, Hauck G, Spann W (1969) Röntgenfluoreszenzanalytische Bleibestimmung bei bekannten Schußentfernungen. Arch Krimin 144: 92
Lima AP de (1964) Les réactions de la diphénylamine et de la brucine dans les récherches de poudre. Ann méd lég 44: 327
Lisicyn AF (1968) Gerichtsmedizinische Begutachtung bei Verletzungen durch Jagdwaffen. Izdatelstvo Medicina Moskau
Lisitsin AF (1963) Determination of the firing distance by the density of pellets in the wound (in Russian). Sudebnomed eksp 6 Nr. 2: 8
Lisitsin AF, Lopatyev AA (1976) Dynamics of powder gases dispersion and its importance in determination of shot distance (in Russian). Sudebnomed eksp 19 Nr. 2: 25

Lisowski Z, Baran E (1968) The securing of powder traces in gun-shot injuries with sprayed lacquer (in Polish with an English Summary). Arch med sadowey 18: 139
Lobanov VI, Gureev AS, Chmir ON (1964) Examination of gun powder grains in medicolegal practice (in Russian). Sudebnomed eksp 7: 7
Lopatenok AA (1961) Imprints on the skin after a shot from a Makarov pistol (in Russian). Sudebnomed Eksp 4: 51
Mach MH, Pallos A, Jons PF (1978) Feasibility of gunshot residue detection via its organic constituents... (Gaschrom.a.Mass Spectr.) J Forens Sci 23: 433
Maiti PC (1973) Powder pattern around bullet holes in blood-stained articles. J Forens Sci Soc 13: 197
Manczarski St (1937) Über die Anwendung der IR-Photographie zur Unterscheidung des Ein- und Ausschusses bei Fernschüssen. Dtsch Z gerichtl Med 28: 366
Marchenko MI, Naynis IVI, Zakaras AP (1978) Injuries to clothings material, skin and cadaveric bones by discharges from 5,45 mm AKS-74 submashine gun (in Russian): Sudebnomed Eksp 21 Nr 2: 17
Marcinkowski T (1963) The characteristics of a shot from short distance on the surface of bones, using sport-type small caliber arms. Arch Med Sadowej 15: 141
Marcinkowski T (1963) L'action thermique d'un coup de feu par fusil sportif de petit calibre. Ann Méd Lég 43: 461
Marcinkowski T (1964) Über die Abhängigkeit der Geschoßdeformation von der Schußentfernung bei Kleinkaliber-Sportwaffen. Dtsch Z gerichtl Med 55: 33
Marcinkowski T (1963) Größe der Einschußöffnung im Knochen und die Schußweite (bei keinkalibriger Sportwaffe). Dtsch Z gerichtl Med 54: 249
Markert K, Flachewski J, Geisler M, Wunderlich G (1971) Erfahrungen mit der Neutronenaktivierungsanalyse bei der Schußentfernungsbestimmung. Kriminal Forens Wiss Nr 4: 199
Matricardi VR, Kilty J (1977) Detection of gunshot residue particles from the hands of a shooter. J Forens Sci 22: 725
Mattoo BN, Nabar BS (1969) Evaluation of effective shot dispersion in buckshot pattern. J Forens Sci 14: 263
Maurer H (1960) Zwei bemerkenswerte Stanzmarken durch Dreyse-Pistolen. Arch Kriminol 125: 24
Mayer FX, Wölkart N (1955) Neue Methode zur Untersuchung von Nahschußspuren. Arch Kriminol 116: 73
Menke H et al. (1972) Determination of gunshot residues by neutron activation analysis. 2nd Intern Conf on Forens Act Anal Glasgow
Menzies RC, Scroggie RJ, Labowitz DI (1980) Characteristics of silenced firearms and their wounding effects. Med Sci Law 20: 201
Messler HR, Armstrong WR (1978) Bullet residue as distinguished from powder pattern. J Forens Sci 23: 687
Molchanov VI, Kalmykov KN, Ozeretskovsky LB (1986) Damage of clothes by shot from close-range distance using AKC 74 Y 5.45 mm shortened submachine gun. Sudebnomed Eksp (Mosk) 29: 22
Moreau TS, Nickels ML, Wray JL, Bottemiller (1985) Pellet patterns fired by sawed-Off shotguns. J Forens Sci 30: 137
Movshovich AA (1966) A feature of a close gun shot from a distant shot (in Russian). Sudebnomed. eksp. 9, Nr 4: 7
Movshovich AA (1971) LM Edlin's polymetallic indicator paper: a tool in the examination of injuries inflicted by firearms. Sudebnomed Eksp 14, Nr 3: 10
Müller F (1939) Bestimmung der Schußentfernung bei Faustfeuerwaffen und Sinoxidmunition. Arch Krimin 104: 142
Nag NK, Mazumdar M (1975) Detection of firearm discharge residues in blood-stained articles by fluorescence. Forens Sci 5: 69
Nag NK, Lahiri A (1986) An evaluation of distribution of pellets due to shotgun discharge. Forens Sci Internat 32: 151
Nasilowski W, Szczepanski J (1968) Anwendung von mikrochemischen und Kontaktmethoden bei Untersuchung von Schußwunden. Aktuelle Fragen Gerichtl Med 3: 306
Nennstiel R, Grooss KD (1980) Zur Beurteilung verkürzter Schrotwaffen. Arch Krimin 168: 76
Nesbitt RS, Wessel JE, Wolten GM, Jones PF Evaluation of a photoluminescence technique for the detection of gunshot residue. J Forens Sci 22: 288

Neuninger H, Bürger H (1969) Beitrag zur chemischen Schußentfernungsbestimmung. Beitr Gerichtl Med 26: 115
Ogarkov IF (1954) Der Einfluß einiger Arten von Kleidung und Schuhwerk auf das Bild der Ein-und Ausschußöffnungen. Probl d Gerichtsmed Expert Moskau S 40
Ordog GJ, Wasserberger J, Balasubramaniam S (1988) Shotgun wound ballistics. J Trauma 28: 624
Palmieri L, Pilleri M (1980) Sugli effetti lesivi di colpi di armi da fuoco corte(revolvers) a proiettile multipli. Prove spe. Zacchia 55: 261
Paradies LH, Gregory CF (1966) The early treatment of close-range shotgun wounds to the extremities. J Bone Joint Surg 48-A: 435
Petersohn F (1969) Experimentelle Untersuchung zur Entfernungsbestimmung bei Schrotschüssen. Beitr Gerichtl Med 26: 117
Pierucci G, Borra G (1970) Cartucce da caccia munite di particolare borra in plastica; rilievi balistici e medicolegali. Arch Soc Lombarda Med Leg 6: 113
Pillay KKS, Driscoll DC, Jester WA (1975) Distribution patterns of firearm discharge residues as revealed by neutron activation analysis. J Radioanalyt Chem 27: 421
Pillay KKS, Thomas CC Jr, Hart DM et al. (1970) Applications of rare earth tracers to gunpowder residues. Nucl Appl Technol 8: 73
Popielski B (1956) Establishing of the distance of shot in the case of serial shots from a automatic weapon (poln.) Arch Med Sadowej 8: 102 (engl. Zsfass)
Pro MJ (1970) Detection of firearm discharge residue by neutron activation analysis—a review. Rev Intern de Pol Crim Nr 242: 270
Prokop O (1960) Ein Beitrag zur Frage der Form der Platzwunde beim absoluten Nahschuß. Arch Krimin 125: 77
Ravreby M (1982) Analysis of long-range bullet entrance holes by atomic absorption spectrophotometry and scanning electron microscopy. J Forens Sci 27: 92
Rios FG, Thornton JI (1984) Discussion of "The application of linear regress, to range-of-fire estimates based on the Spread. J Forens Sci 29: 695
Rolfe HC, Curle D, Simmons D (1971) A histological technique for forensic ballistics. J Forens Med 18: 47
Rowe WF, Hanson SR (1985) Range-of-fire estimates from regression analysis applied to the spread of shotgun pellet pattern. Forens Sci Intern 28: 239
Ruch RR, Guinn VP, Pinker RH (1964) Detection of gunpowder residues by neutron activation analysis. Nucl Sci and Engin 20: 381
Rudzits E, Wahlgren M (1975) Firearm residue detection by instrumental neutron activation analysis. J Forens Sci 20: 119
Rudzitis E, Kopina M, Wahlgren M (1973) Optimization of firearm residue detection by neutron activation analysis. J Forens Sci 18: 93
Ryvarden G (1976) Problematik der Schußentfernungsbestimmung mit dem REM bei Distanzen über 1 m. Beitr Gerichtl med 24: 179
Schelemann B, Wüst E (1983) Verletzungsmöglichkeiten im Brustraum durch Druckluftwaffen—Rekonstruktion einer fraglichen Selbsttötung. Beitr Gerichtl Med 41: 75
Schmechta H, Eigendorf HG (1980) Asymmetrisches Schmauchbild nach Veränderung der Waffenmündung der Maschinenpistole "Kalaschnikow". Kriminal Forens Wiss Nr 39: 47
Schmechta H, Weinke H (1970) Nahschußzeichen bei baugewerblichen Bolzenschußgeräten. Z Militärmed 11: 346
Schmechta H, Ziegler M (1973) Gesichtspunkte zur Schußentfernungsbestimmung bei Mehrfachschüssen aus der Maschinenpistole "Kalaschnikov". Z Militärmed 3: 140
Schmechta H, Ziegler M, Weinke H (1972) Studien zur Schußentfernungsbestimmung bei Schüssen aus baugewerblichen bolzenschußgeräten. Arch Kriminol 149: 11
Schmitz J (1973) Erfahrungen mit der Methode der Autoradiografie zur Bestimmung der Schußentfernung. Kriminalistik 27: 207
Schmitz J (1973) Experience with the method of autoradiography for firing distance determination. J Radioanalyt Chem 15: 219
Schöntag A (1957) Vorschlag einer neuen Methode: Bestimmung der Schußentfernung mittels des "Schmauchringes" Arch Krimin 120: 62
Schöntag A (1964) Die Entwicklung der Methode zur Bestimmung der Schußentfernung und ihre historischen Hintergründe. Arch Krimin 134: 137
Schöntag A (1970) Bestimmung der Schußentfernung auf Grund besonderer Anschmauchungen. Arch Krimin 144: 152

Schöntag A (1971) Steigerung des Meßbereichs für spektrographische Schußentfernungsbestimmung durch Kaltveraschung des Textilmaterials. Arch Krimin 147: 32
Schöntag A, Baumgärtner F (1963) Erweiterung des Meßbereichs auf 3 m bei der Bestimmung der Schußentfernung durch Anwendung der Aktivierungs Analyse. Arch Krimin 131: 1
Schöntag A, Heindl R (1956) Entwicklung der Methoden zur Bestimmung der Schußentfernung. Arch Krimin 118: 19
Schöntag A, Suchenwirth H (1970) Eine neue Methode zur Bestimmung kleiner Schußentfernungen zwischen 0,5 und 5 cm. Arch Krimin 146: 61
Schöntag A (1957) Neue Methode: Bestimmung der Schußentfernung durch spektrographische Spurenanalyse der "Schmauchelemente" Antimon, Blei oder Barium. Arch Krimin 120: 4
Schulz G, Schewe G, Schütz H et al. (1981) Untersuchungen von Gaspistolenschüssen. Arch Krimin 168: 165
Schweppe H (1969) Neuer Komplexbildner zum Nachweis von Spuren Kobalt, Kupfer, Zink and Cadmium. Z Anal Chem 244: 312
Scott HD et al. (1966) Investigation of firearm discharge residues AWRE-Report No. 0–5/66 (UK Atom En Auth)
Seamster A, Mead T, Gislason J, Jackson et al. (1976) Studies of the spatial Distribution of Firearms Discharge Residues. J Forens Sci 21: 868
Sellier K (1965) Verteilung der Elemente Pb und Sb im Schmauch und Schußentfernungsbestimmung. Dtsch Z Gerichtl Med 56: 39
Sellier K (1967/1988) Schußentfernungsbestimmung. Schmidt-Römhild, Lübeck 1967, 2nd Edn 1988
Sellier K (1968) Erweiterung des Schußentfernungsbereiches durch Anreicherung der Schmauchelemente. Arch Krimin 141: 34
Sellier K (1968) Schußentfernungsbestimmung mit Hilfe von Elementen, die aus dem Geschoßmantel stammen. Arch Krimin 141: 73
Sellier K (1969) Schußwaffen und Schußwirkungen. Schmidt-Römhild, Lübeck
Sellier K (1977) Schußwaffen und Schußwirkungen Vol. II Schmidt-Römhild. Lübeck
Sellier K (1982) Schußwaffen und Schußwirkungen Vol. I. Schmidt-Römhild, Lübeck, 2nd edn
Sen P, Panigrahi N, Rao MS et al. (1982) Application of proton-induced X-ray emission technique to gunshot residue analysis. J Forens Sci 27: 330
Sigrist Th (1983) Über die Entstehung der Oberhautverletzung am Einschuß beim Schuß mit aufgesetzter Waffe. Z Rechtsmed 93: 199
Sigrist T und Dirnhofer R (1980) Zur Morphologie des Einschusses bei Kleinkaliberschüssen. Ein Beitrag zur Entstehung der Stanzverletzung. Beitr Gerichtl Med 38: 101
Simon A (1967) Die Bedeutung der sog. Hautspaltlinien bei der Entstehung der Platzwunde beim absoluten Nahschuß. Akt Fragen Gerichtl Med 2: 67
Sinatra S (1982) Diagnosi istologica di entrata e di distanza di sparo in un caso con lunga Sopravvivenza. Med Leg Assicur 4., suppl., 243
Sinha JK, Misra GJ (1971) Detection of powder particles at the crime scene. J Forens Sci 16: 109
Skopin IV Über die Möglichkeit der Methodik der Erkennung des Pulverrußes auf dunklen Geweben mittels Untersuchung. Probl Gerichtsmed Expertise, S. 82–84, Moskau
Skopin IV (1954) Zur Frage der Bestimmung des Nahschußabstandes nach der Ablagerung von Pulverruß um die Einschußöffnung. Probl. Gerichtsmed Expertise, S. 65–70, Moskau
Smith C (1973) Spherical coordinates for plotting the positions of powder residues. J Forens Sci 18: 101
Smusin JS (1954) Zur Frage der Bestimmung des Schießabstandes aus Schrotflinten. Probl Gerichtsmed Expert S 101 Mosaku
Smusin JS (1954) Bestimmung von Schußabstand und -winkel aus einer Schrotflinte auf rechnerischem Wege. Probl Gerichtsmed Expert S 109 Mosakau
Speak RD, Kerr FC, Rowe WF (1985) Effects of range, caliber, barrel length and rifling on pellet patterns produced by shotshells. J Forens Sci 30: 412
Spitz WU (1969) Rikochett- oder Kontaktschußverletzung. Dtsch Z Gerichtl Med 66: 153
Steinberg M, Leist Y, Tassu M (1984) A new field kit for bullet hole identification. J Forens Sci 29: 169
Stephens BG, Allen TB (1983) Back spatter of blood from gunshot wounds—observations and experimental simulation. J Forens Sci 28: 437
Stone IC (1987) Observations and statistics relating to suicide weapons. J Forens Sci 32: 711
Stone IC, Dimaio VJM (1977) Metallic Residues in Gunshot Wounds. J Assoc of Firearms a Toolm Examiners May
Stone IC, Petty CS (1974) Examination of gunshot residues. J Forens Sci 19: 784

Stone IC, Dimaio VJM, Petty CS (1978) Gunshot wounds: visual and analytical procedures. J Forens Sci 23: 361
Suchenwirth H (1972) Neue Methode zur Schußentfernungsbestimmung. Die Neue Polizei ##, 26
Suchenwirth H (1972) Ein einfaches spezifisches Abdruckverfahren zum Erfassen und Beurteilen von Schmauchbildern. Arch Krimin 150: 152
Tardaguila RF, Piga Rivero A (1975) Lesiones producidas por proyectiles, impulsados por armos de aire o de otro gas comprimido. Rev Esp Med Leg 2: 24
Tassa M, Leist Y, Steinberg M (1982) Characterisation of gunshot residues by X-ray diffraction. J Forens Sci 27: 677
Tassa M, Adan N, Zeldes N, Leist Y (1982) A field kit for sampling gunshot residues particles. J Forens Sci 27: 671
Tewari SN (1973) Ein neues mikrochemisches Reagenz zum Nachweis von Schießpulverrückständen in Fällen von Schußwaffengebrauch. Arch Krimin 152: 96
Thieme H (1984) Schußentfernungsbestimmung bei relativen Nahschüssen mit Hilfe der energiedispersiven Röntgenfluoreszensanalyse. Krim u Forens Wiss H 53/54, 94
Thieme H (1986) Schußentfernungsbestimmung ohne Tatwaffe. Krim u Forens Wissensch H 61, 62: 58
Thornton JI (1986) Close Proximity Gunshot Residues. J Forens Sci 31: 756
Thornton JI, Crim D, Rios FG (1987) The effects of deformed shot on shotgun pellet dispersion. J Forens Sci 32: 522
Tsykunov AP (1971) Characteristics of the injuries caused by stud-gun; an experimental study. Sudebnomed Eksp 14: 35
Tzykunov AP (1969) Characteristics in distribution of a close-shot features produced by discharge of special pistol applied in building technique. Exper study Sudebnomed Eksp 12 Nr 4: 32
Viksnin YS, Pokrovsky EN (1962) An unusual sign of firing point blank. (in Russian) Sudebnomed Eks Nr 4: 52
Villanueva E. Hernandez-Cueto C, Lorente JA et al. (1987) Study of inorganic residues around entrance and exit projectile orifices. J Forens Sci 32: 1079
Vinogradov IV (1954) Einige Eigentümlichkeiten des morphologischen Bildes der Schußöffnung in der Haut bei Schüssen in diese durch... Kleidung aus weiterm Abstand. Probl Gerichtsmed Expertise S 71 Moskau
Walker JT (1937) New Engl. J Med 216: 1024
Wand A (1938) Beeinflussung von Schußspuren der Sinoxidmunition durch äußere Einwirkungen (Experimentelle Untersuchung an ausgeschnittenen Hautstücken) ID Göttingen
Weber W (1981) Streifschußverletzungen. Z Rechtsmed 87: 35
Weigel B, Wolfram D, Lindner R et al. (1981) Suicid durch Genickschußverletzung. Kriminal Forens Wiss Nr 43: 73
Welz B (1983) Atomabsorptionsspektrometrie. Verlag Chemie Weinheim, 3rd edn
White RS, Owens AD (1987) Automation of gunshot residue detection and analysis by SEM/EDX. J Forens Sci 32: 1595
Wickenhaüser K (1939) Über die Möglichkeiten eines quantitativen und qualitativen Nachweises im SchuBfeld bei Verwendung der Sinoxidmunition. Dtsch Z Gerichtl Med 31: 298
Wojahn H (1968) CO-Hb-Konzentration im Schußkanal als Zeichen des Nahschusses. Beitr Gerichtl Med 24: 190
Wolten GM, Nesbitt RS, Calloway AR et al. (1979) Particle analysis for the detection of gunshot residues. I. Scanning electron microscopy/energy dispersive X-ray characterisation of hand deposits. J Forens Sci 24: 409
Wolten GM, Nesbitt RS (1980) On the mechanism of gunshot residue particle formation. J Forens Sci 25: 533
Wray JL, McNeil JE, Rowe WF (1983) Comparison of methods for estimating range of fire based on the spread of buckshot pattern. J Forens Sci 28: 846
Wyttenbach A (1971) Anwendung der Aktivierungsanalyse in den forensischen Wissenschaften. Beitr Gerichtl Med 28: 326
Zaki SA, Hanzlick R (1987) Gunshot wound with asphalt related pseudo-shot, pseudo-tattooing, and pseudo-scorching. J Forens Sci 32: 1136
Zeichner A, Foner HA, Dvorachek M, Bergman et al. (1989) Concentration Techniques for the Detection of gunshot residues by scanning electron microscopy/energy dispersive X-ray analysis (SED/EDX). J Forens Sci 34: 312
Ziegler M (1965) Zur Einsatzmöglichkeit der Polarografie in der Schußexpertise. Aktuelle Fragen d Gerichtl Med Halle

Ziegler M (1968) Neue Ergebnisse zum Problem der polarografischen Bestimmung der Schußentfernung. Aktuelle Fragen d Gerichtl Med 3: 169

Ziegler M (1981) Schußentfernungsbestimmung bei ungewöhnlicher Waffe, ein Beitrag zur Genauigkeit der Methodik. Kriminal Forens Wiss Nr 42: 49

Ziegler M, Fehse B (1966) Schußentfernungsbestimmung mit dem Tastpolargrafen. Arch Krimin 137: 71

Ziegler M, Schmechta H (1970) Zur Bestimmung der Schußentfernung bei Schüssen aus der Maschinenpistole "Kalaschnikow". Z Militärmed 5: 285

Ziegler M, Stock L (1971) Einige Gedanken zur Schußentfernungsbestimmung bei Schrotschüssen. Kriminal Forens Wiss Nr 4: 205

Zumwalt RE, Campbell B, Balraj E et al. (1981) Wounding characteristics of "shotshell" ammunition: report of a case. J Forens Sci 26: 198

Zwicker H (1966) Untersuchungen über die Verteilung des Schmauchs beim Nahschuß und der Anwendung zur Schußentfernungsbestimmung. ID Bonn

Subject Index

Author Index Volumes 1–6

The volume numbers are printed in italics

MIX
Papier aus verantwortungsvollen Quellen
Paper from responsible sources
FSC® C105338

If you have any concerns about our products,
you can contact us on
ProductSafety@springernature.com

In case Publisher is established outside the EU,
the EU authorized representative is:
Springer Nature Customer Service Center GmbH
Europaplatz 3, 69115 Heidelberg, Germany

Printed by Libri Plureos GmbH
in Hamburg, Germany